NURTURING REFLECTIVE LEARNERS IN MATHEMATICS

Yearbook 2013

Association of Mathematics Educators

NURTURING REFLECTIVE LEARNERS IN MATHEMATICS

Yearbook 2013
Association of Mathematics Educators

editor

Berinderjeet Kaur

National Institute of Education
Nanyang Technological University
Singapore

Published by

World Scientific Publishing Co. Pte. Ltd.
5 Toh Tuck Link, Singapore 596224
USA office: 27 Warren Street, Suite 401-402, Hackensack, NJ 07601
UK office: 57 Shelton Street, Covent Garden, London WC2H 9HE

British Library Cataloguing-in-Publication Data
A catalogue record for this book is available from the British Library.

NURTURING REFLECTIVE LEARNERS IN MATHEMATICS
Yearbook 2013, Association of Mathematics Educators

ISBN 978-981-4472-74-6

Printed in Singapore by World Scientific Printers.

Contents

Chapter 1

Nurturing Reflective Learners in Mathematics: An Introduction

Berinderjeet KAUR

This introductory chapter provides an overview of the chapters in the book. The chapters are organised according to three broad themes: fundamentals, instructional tools, and approaches to teaching for nurturing reflective learners in mathematics classrooms. It ends with some concluding thoughts that readers may want to be cognizant of while reading the book and also using it for reference and further work.

1 Introduction

This yearbook of the Association of Mathematics Educators (AME) in Singapore focuses on Nurturing Reflective Learners in Mathematics. Like three of our past yearbooks, Mathematical Problem Solving (Kaur, Yeap, & Kapur, 2009), Mathematical Applications and Modelling (Kaur & Dindyal, 2010), and Reasoning, Communication and Connections in Mathematics (Kaur & Toh, 2012) the theme of this book is also shaped by the framework, shown in Figure 1, of the school mathematics curriculum in Singapore. The primary goal of school mathematics in Singapore is mathematical problem solving and amongst the five inter-related components that the framework places emphasis on is metacognition. In elaborating the framework, for both the primary and secondary students, the Ministry of Education (MOE) (2012a, 2012b), syllabus documents clarify that:

Metacognition, or thinking about thinking, refers to the awareness of, and the ability to control one's thinking processes, in particular the selection and use of problem-solving strategies. It includes monitoring of one's own thinking, and self-regulation of learning. To develop metacognitive awareness and strategies, and know when and how to use the strategies, students should have opportunities to solve non-routine and open-ended problems, to discuss their solutions, to think aloud and reflect on what they are doing, and to keep track of how things are going and make changes when necessary (Ministry of Education, 2012a, p. 17; 2012b, p. 16).

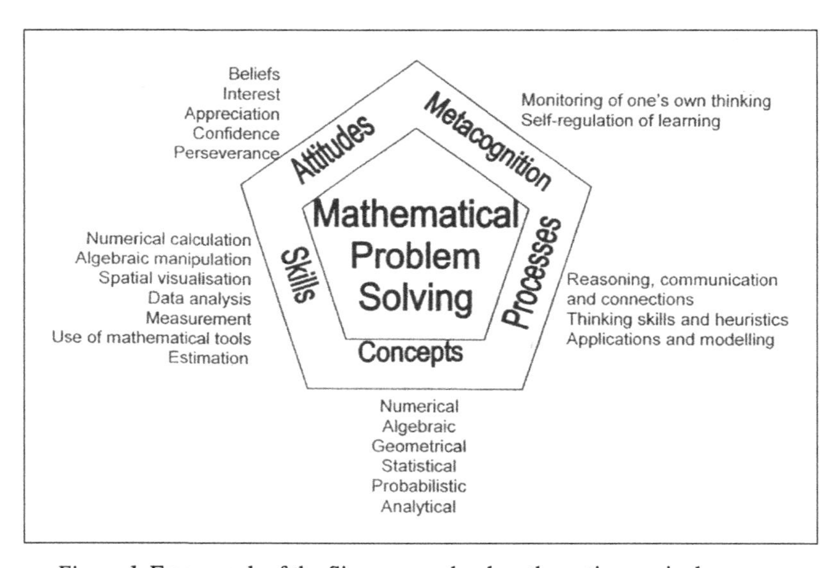

Figure 1. Framework of the Singapore school mathematics curriculum

In addition the syllabus documents outline three principles of mathematics teaching and three phases of mathematics learning in the classrooms. The three principles of teaching are as follows:

Principle 1 — Teaching is for learning; learning is for understanding; understanding is for reasoning and applying and, ultimately problem solving.

Principle 2 — Teaching should build on students' knowledge; take cognizance of students' interests and experiences; and engage them in active and ***reflective learning***.
Principle 3 — Teaching should connect learning to the real world, harness ICT tools and emphasise 21st century competencies (Ministry of Education, 2012a, p. 21; 2012b, p. 23).

The three phases of mathematics learning in the classrooms are as follows:

Phase I — Readiness
Student readiness to learn is vital to learning success. In the readiness phase of learning, teachers prepare students so that they are ready to learn. This requires considerations of prior knowledge, motivating contexts, and learning environment.

Phase II — Engagement
This is the main phase of learning where teachers use a repertoire of pedagogies to engage students in learning new concepts and skills. Three pedagogical approaches, activity-based learning, teacher-directed inquiry, and direct instruction, form the spine that supports most of the mathematics instruction in the classroom. They are not mutually exclusive and could be used in different parts of a lesson or unit. For example, the lesson or unit could start with an activity, followed by teacher-led inquiry and end with direct instruction.

Phase III — Mastery
This is the final phase of learning where teachers help students consolidate and extend their learning. The mastery approaches include: motivated practice, ***reflective review*** and extended learning. (Ministry of Education, 2012a, pp. 22-25; 2012b, pp. 24-27).

The document further expands ***reflective review*** and elaborates that "It is important that students consolidate and deepen their learning through tasks that allow them to ***reflect on their learning***. This is a good habit that needs to be cultivated from an early age and it ***supports the development of metacognition***" (Ministry of Education, 2012a, p. 25; 2012b, p. 27).

The framework, learning principles and phases of learning in the Singapore school mathematics syllabus documents signal very clearly the need for teachers to engage their students in reflecting about their actions before, during or after solving mathematical tasks. Such practice will allow students to create habits of the mind that lead to monitoring their own thinking and also regulating their own learning. According to Kriewaldt (2001), 'reflection is an integral element of metacognition as it is the means by which one monitors thinking processes' (p. 3). In addition, reflection often contains elements of self-regulation that are stimulated through questions such as "how would I do this differently the next time?"

In reviewing Dewey's work, Rodgers (2002) states that, "reflection is a particular way of thinking and cannot be equated with mere haphazard 'mulling' over something" (p. 849). The importance of reflecting on 'something' is also emphasised by Wheatley (1992), who argues that "reflection plays a critically important role in mathematics learning and that just completing tasks is insufficient" (p. 529). In attempting to address the challenge of how teachers, in Singapore schools, may nurture reflective learners in their mathematics classrooms the theme of the 2011 conference for teachers, jointly organised by the Association of Mathematics Educators (AME) and the Singapore Mathematical Society (SMS), was appropriately *Nurturing Reflective Learners*.

The following 14 peer-reviewed chapters resulted from the keynote and invited lectures delivered during the conference. The authors of the chapters were asked to focus on evidence-based practices that school teachers can experiment in their lessons to bring about meaningful learning outcomes. The chapters are categorised into three main sections, namely the fundamentals, instructional tools, and approaches to teaching for nurturing reflective learners in mathematics classrooms. It must be noted that the 14 chapters do give a reader some ideas about the why, what and how of nurturing reflective learners in mathematics lessons. However, in no way are the 14 chapters a collection of all the know-how of the subject.

2 Fundamentals for Nurturing Reflective Learners

Learning results from students' active participation in activities. In the classroom, teachers guide students in their learning through instructional activities. Often teacher's knowledge and beliefs guide them in shaping the activities. So, in attempting to nurture reflective learners in mathematics a few fundamentals necessary are explored in this book. They are the cognitive aspect of reflection, engagement of the whole psyche in aspects of reflection, knowledge and beliefs of teachers necessary for the development of reflective learners, metacognitive reflection and education of teachers for advancing reflection in their classrooms. It is befitting that in chapter 2, Voon explains in simple language the cognitive aspect of reflection that is the basis of learning and mastery from the perspective of a professor who teaches medical and dental students, doctors, dentists, obstetricians, psychiatrists and surgeons at the National University of Singapore. He advocates the habitual practice of reflection, an intellectual exploration of the various pathways of thought using our conscious mind, to form and strengthen new neural pathways. Mason in the next chapter, draws the attention of the readers for the need of working with the whole psyche when nurturing reflective learners. He notes that the whole person is an intricate interweaving of intellect-cognition, emotion-affect and behaviour-enaction with attention and will. Therefore by working on some mathematics together and reflecting on actions, one can learn from the experience of doing the mathematics. This is often done by withdrawing from the activity at some point and adopting a reflexive stance, asking oneself questions such as 'what was ineffective and what was effective? ......what did I learn about myself (dispositions, propensities, habits, etc...)?'

Beswick in chapter 4, highlights that both teachers' knowledge and beliefs are crucial to the development of reflective learners. Using rational number concepts as a basis she draws on learning episodes to illustrate behaviours of reflective learners and teacher actions that appear to support reflective learning and the teacher knowledge and beliefs that appear to underpin these. In chapter 5, Wong coins the word metacognitive reflection to include elements of metacognition (such as awareness and regulation) and reflection (as in looking back in the

Polya's model of problem solving) to examine roles for mathematics instruction at secondary level. He examines two aspects of metacognitive reflection: metacognition during problem solving and regulation of learning. He reviews local studies and also draws on findings from literature thus outlining approaches for teaching metacognition and regulation of learning.

To nurture reflective learners, teachers must be reflective learners themselves. Hence, as part of teacher education teachers must have opportunities to engage in reflection that may take any of the forms: refection-in-action (minute by minute decisions a teacher may make on his or her actions when executing a lesson), reflection-on-action (looking back at a lesson that has passed) and reflection-for-action (desired outcomes guiding the plan for a lesson) (Schön, 1983; Killion & Todnem, 1991 cited in Harford, MacRuairc, & McCartan, 2010). Lim and Chew, in chapter 6, demonstrate how they used a video recorded lesson of a primary mathematics excellent teacher to promote reflective thinking among in-service teachers in Malaysia and Singapore. Lastly in the next chapter, Kissane claims that good teachers are often described as 'reflective practitioners' and that nurturing reflective learning in the school depends critically on teachers being appropriately reflective themselves. He describes in the chapter some experiences that were designed to encourage pre-service teachers to reflect on aspects of their learning and consider what their teacher educator might learn from it also.

3 Instructional Tools for Nurturing Reflective Learners

In the context of this book, instructional tools are limited to basically mathematical tasks and a teacher designed strategy that may be used to engage students in reflecting about their work in the mathematics classroom.

It is apparent that central to all mathematics lessons are mathematical tasks. A mathematical task is defined as a set of problems or a single complex problem that focuses students' attention on a particular mathematical idea (Stein, Grover, & Henningsen, 1996). From the

TIMSS Video Study (NCES, 2003), in which Australia, Czech Republic, Hong Kong, Japan, Netherlands, Switzerland, and the United States participated, it was found that students spent over 80% of their time in mathematics class working on mathematical tasks. According to Doyle (1988), "the work students do, defined in large measure by the tasks teachers assign, determines how they think about a curricular domain and come to understand its meaning" (p. 167). Hence different kinds of tasks lead to different types of instruction, which subsequently lead to different opportunities for student learning (Doyle, 1988).

Watson, in chapter 8, shows that reflection on a sequence of work can enable learners to recognise similarities at a higher, more abstract, level than they had experienced while doing the work. She illustrates with examples of routine calculation practice how teachers can engage students to reflect on the effects of actions and uncover underlying relations. In chapter 9, Kemp constructs a framework comprising six criteria that teachers may use to craft suitable mathematical tasks for engaging students in reflective thinking. The six criteria for tasks are: be unfamiliar or non-routine to the students, have different possible approaches or methods of solution, require some thinking about constraints or assumptions, be appropriate to the level of mathematical knowledge and experience of the students, be potentially interesting and motivating to the students; and require justification or explanation of the reasons for the chosen approach and solution. She illustrates through three examples the use of the framework and interpretations of the six criteria. In the following chapter, Toh draws on mathematics competition questions and illustrate how teachers may adapt them for students to work and reflect on, thereby engaging in higher order thinking and also challenging them to re-examine their originally acquired mathematical notions and beliefs. He draws on several algebra competition questions to illustrate how appropriate tasks for classroom use may be crafted. In chapter 11, Kwon and Lee focus on an aspect of reflective thinking, i.e. looking back on one's problem solving and thinking to analyse mathematical tasks on the topic: Polar coordinates in advanced mathematics. These tasks are suitable for nurturing reflective thinking amongst high school students of mathematics.

The last chapter in this category shows an attempt by a primary school teacher to devise a strategy for engaging students to reflect on their errors in mathematics written work. Wong, in chapter 12, shows how he adopted an idea used mainly by English Language teachers in Singapore to grade written compositions and give their students feedback for use in his mathematics lessons. He devised codes and used them to give his students feedback about their errors when grading their written work.

4 Approaches to Teaching for Nurturing Reflective Learners

A lesson that has as its focus dissemination of knowledge by the teacher would certainly have little or no scope for teacher questions or knowledge construction by students. In contrast, a lesson that has as its focus the development of reflective attributes amongst students would involve significant student participation in the lesson. The three chapters in this category each illustrate a variation of the types of teacher actions that promote reflective learning. Hino, in chapter 13, shows the dominant presence of "reflective learning" in mathematics lessons in Japan. This is verified by the vocabulary that exists to describe teacher actions. In the chapter she describes the Japanese problem solving approach and several perspectives for enhancing reflective thinking and learning in classroom practice of mathematics lessons. She also draws on some empirical data to show that there are a variety of ways through which students may be stimulated to reflect on their learning. In chapter 14, Koyama states that we need to look into students' process of understanding mathematics for insights that may help us to think of ways to nurture reflective learners. He analyses a mathematics lesson in Japan and arrives at four suggestions for teachers who aim to nurture reflective learners in their classrooms. The suggestions are firstly teachers should pay attention not only to the correctness of student answers to mathematical tasks but also to the processes that lead to the answers. Secondly, it is important for teachers to make a decision on the kinds of learning situation that need to be set up so as to help students improve their mathematical understanding. Thirdly, the social interaction between teacher and students is important to ensure a high level of mathematical understanding. Lastly, students should be encouraged to reflect on what they have done and also do adequate activities to integrate their knowledge.

A framework for learning and teaching can guide teachers in planning their lessons to enact specific goals. In the last chapter of the book, Limjap shares with readers one such framework, the Learner Centred Learning Environment framework, comprising four phases that is used in an elementary school in the Philippines. The first phase, identifying the students' prior knowledge, is where both teachers and students have opportunities to engage in reflective thinking. In this phase teachers should reflect on how best to elicit the varied conceptions of students in creative ways. Teachers should also guide students to reflect on their understanding and past learning. In the chapter Limjap draws on a mathematics lesson, in grade 4 and shows how the teacher nurtured reflective thinking in her classroom discourse and students' seatwork.

5 Some Concluding Thoughts

Wong and Lee (2010) in their evaluation of the framework, shown in Figure 1, which has been the in place since 1990 noted that:

> Of the five factors, metacognition is the most problematic one because many teachers find it very difficult to incorporate metacognition into their mathematics lessons (p. 99).

As teaching activities to inculcate metacognition include discussing alternative solutions, thinking aloud, following Polya's four steps, and monitoring through self-questioning (Wong, 2002) teachers need to strike a balance in their classroom pedagogies between meaningful learning and also preparation for examinations. For this to occur, teachers need to move away from their main preoccupation with drill and practice as observed by Wong and Lee (2010):

> Drilling students to solve standard problems is a major part of mathematics instruction in Singapore and many other countries.

Singapore teachers often construct exercises that parallel past examination questions, which have been compiled and sold to the public as Ten-Year Examination Series. Even problems with initially unfamiliar contexts or "challenge problems" gradually become routine to students with frequent exposure (p. 102).

The chapters in the yearbook provide readers with some ideas on the why, what and how of nurturing reflective leaners. In the revised school mathematics curriculum explicit emphasis has been placed on reflection as a vehicle to develop metacognition (see p. 3). It must be noted that engagement in reflection nurtures metacognition thereby developing young adults with habits of the mind to monitor their own thinking and also regulate their own learning. Lastly, a word of caution for readers who are often overwhelmed by yet more ideas in the book in view of their already content-heavy curriculum. We urge the readers to read the chapters carefully and try some of the ideas in their classrooms and convince themselves that these ideas offer a means of infusing metacognition in their lessons and engage students in meaningful mathematical practices.

References

Doyle, W. (1988). Work in mathematics classes: The context of students' thinking during instruction. *Educational Psychologist, 23*, 167-180.

Harford, J., MacRuairc, G., & McCartan, D. (2010). 'Lights, camera, reflection': Using peer video to promote reflective dialogue among student teachers. *Teacher Development, 14*(1), 57-68.

Kaur, B., & Dindyal, J. (2010). *Mathematical applications and modelling*. Singapore: World Scientific.

Kaur, B., & Toh, T.L. (2012). *Reasoning, communication and connections in mathematics*. Singapore: World Scientific.

Kaur, B., Yeap, B.H., & Kapur, M. (2009). *Mathematical problem solving*. Singapore: World Scientific.

Kriewaldt, J. (2001). A thinking geography curriculum. *Interaction, 29*(4), 1-7.

Ministry of Education, Singapore (2012a). *O-Level, N(A) Level, N(T) level mathematics teaching and learning syllabuses*. Singapore: Author.

Ministry of Education, Singapore (2012b). *Primary mathematics teaching and learning syllabus*. Singapore: Author.

NCES (National Center for Educational Statistics). (2003). *Teaching mathematics in seven countries: Results from the TIMSS video study*. Washington, DC: U.S. Department of Education.

Rodgers, C. (2002). Defining reflection: Another look at John Dewey and reflective thinking. *Teachers College Record, 104*(4), 842-866.

Schön, D. (1983). *The reflective practitioner: How professionals think in action*. New York: Basic Books.

Stein, M.K., Grover, B., & Henningsen, M. (1996). Building student capacity for mathematical thinking and reasoning: An analysis of mathematical tasks used in reform classrooms. *American Educational Research Journal, 33*, 455-488.

Wheatley, G.H. (1992). The role of reflection in mathematics learning. *Educational Studies in Mathematics, 23*(5), 529-541.

Wong, K.Y. (2002, April). Helping your students to become metacognitive in mathematics: A decade later. *Mathematics Newsletter, 12*(5), Retrieved on Dec 10, 2012 from http://math.nie.edu.sg/kywong/

Wong, K.Y., & Lee, N.H. (2010). Issues of Singapore mathematics education. In K.S.F. Leung & Y. Li (Eds.), *Reforms and issues in school mathematics in East Asia* (pp. 91-108). Rotterdam, The Netherlands: Sense.

Chapter 2

The Neurocognition of Reflection: The Mystery in Learning, the Essence of Teaching, From Mystery to Mastery

Frank Chee Tet VOON

Recent advances in medical and information technology have provided us with an insightful glimpse into the processes of the mind and the cognitive functions of the brain. Technology is also changing the way we learn, making everyone both a teacher and a student at the same time. Teaching newer generations of students is becoming increasingly multidimensional. These online citizens have grown up in a dominantly technological environment that has undergone rapid change in the 21st century. In addition, both the generation of online data and the ease of anywhere, anytime, mobile cloud computing with 24/7 wi-fi cell phone access to multimedia content has shifted the paradigm from teaching to learning, and from information creation to knowledge curation and distillation. This chapter focuses on a learning paradigm based on new and emerging ideas in neurocognition, to provide us with the knowledge of how we can efficiently and effectively think through data to information to knowledge to mastery, through object definition and the experience of reality. There are ways to develop confidence in students, and one way is the path from understanding to practice coupled along the course with an intellectual appreciation of symbolic representation (pattern formation) and teaching to learn based on question formulation rather than answer routines alone. All these have great implications for both students and teachers of Mathematics, in relating relevant new principles of Innovation,

Numeracy, and Distraction and Differentiation to good practices in the classroom.

1 Introduction

The title and abstract of this chapter follows the theme of this book "Nurturing Reflective Learners", highlighting the importance of metacognition in general and reflection in particular as well as the place of both pedagogy and mathematical content knowledge within the Singapore mathematics curriculum, and their importance to mathematics teachers in our schools. One of the Desired Outcomes of Education (DOE) for pupils in Singapore schools is directed at the development of self-directed learners who are conscious of the responsibility that they should have for their own learning, and who are able to question and reflect upon the knowledge they endeavour to gain as they persevere to understand data and information in their pursuit of personal learning.

Cognitive neuroscience is giving us solid evidence of neural pathways in the brain that are involved in memory formation and therefore in the acquisition of new data, facts and information too. Knowledge of neurocognitive concepts can lead us to an appreciation of two important phases in learning, namely that of the importance of accurate initial understanding followed by repetition and practice in order to consolidate the facts, information and knowledge into long term memory and retention for future recall.

This understanding can now be used by teachers and applied in pedagogical practice in the classroom to enhance efficiently and effectively the learning of each individual student who can therefore progress at his or her own pace. Catering to the individual development of each student will require the use of information technology to overcome the constraints of limited teacher-student contact time in both instruction as well as feedback that includes the correction of homework and relevant guidance when mistakes are made and detected. The creative use and application of multimedia courseware is one such way that can be implemented in the present day at relatively low cost to bring about such individually specific benefits for personal deep practice.

This chapter introduces the topics of networks of neurons, executive function for the processes of decision making and conscious understanding in the prefrontal cortex, the wiring together of nerves in pathways that will strengthen their ability to fire together and the increase in the myelin sheath surrounding the axons of those nerves that will enhance the speed and strength of conduction of nerve impulses. These concepts and details will be illustrated with a real world analogy of neural networks in terms of a physical map of a city, together with examples that can be commonly encountered by teachers today so that readers will begin to understand the underlying neurocognitive principles and be inspired to learn more on their own and apply this in their teaching profession.

2 How Do We Really Learn?

Most people, teachers included, do not really know how they themselves learn something new. They know that the brain is an organ, like the heart or liver or kidney in the human body. Just as the liver is made up of liver cells and there are cardiac muscle cells in the heart, the brain consists of nerve cells or neurons that require supporting cells known as neuroglia to maintain their proper function in addition to the blood vessels that supply the brain.

For our purposes, it will be sufficient to just know a little about two aspects of this in terms of the neuron and its myelin sheath. The neuron is commonly known as the nerve cell. The myelin sheaths are produced by the neuroglia to surround the axons which are processes of the neuron that conduct or transmit nerve impulses. While we consider the brain as an organ, we also know that the mind is a process and that this is the property or function of the brain, just as the function of the heart involves the circulation of blood and the liver has metabolic functions, while the function of the kidney involves the filtration of blood.

However when one actually thinks about how one learns a new topic or subject, one usually bases it on past experience which is to read and read repeatedly until the new names or ideas or concepts finally "sink in" or one at times memorizes repeatedly until one becomes word perfect.

Thus far, there has been no anatomical or organic basis for the periods of study that students have to undergo in mastering a subject. That is the mystery of learning that most students and adults have encountered in their years of schooling.

Is there a way to move from the mystery of learning to the mastery of learning? There is, and it is based on recent findings that researchers in the fields of neurocognition, psychology and computational intelligence have begun to uncover.

3 The Two Phases of Understanding and Recall

The initial phase in learning a topic lies in understanding the relationship of facts that make a meaningful principle or concept that can be applied to make sense of a new word or phrase or a certain collection of data, facts or information. The new facts and details that are taught are first processed in short-term memory and will therefore have to be linked to information that the student already possesses which is stored in his or her long-term memory. This is essential for meaningful retention of knowledge to occur and for subsequent recall from long-term memory.

The subsequent phase is to repeatedly practice this new information, especially the parts that may be difficult to grasp or integrate so as to strengthen its recall. The difficult parts will vary for different learners, and deep practice involves practicing just these parts by each individual rather than a one size fits all approach where the whole group of students do it together at the same time.

4 Neuroanatomy

The nervous system is an intricate network of nerve cells and supporting cells that coordinates and controls all the functions of the body including intellectual processes like cognition which takes place in the frontal lobe of the brain in a region known as the prefrontal cortex. The neuron has basically a cell body that extends into an elongated axon which ends in numerous small branches known as dendrites (Drake, Vogl, & Mitchell, 2010). Through these extensions, a single nerve can connect with up to an estimated 10,000 other neurons. There are thus billions of neural circuits in the functioning brain.

The neuronal cell body has a specific site at the beginning of the axon which can produce an electrical stimulus or nerve impulse that can travel the entire length of the axon to its terminal dendrites. There are various chemicals known as neurotransmitters stored in small containers or vacuoles in the region of the dendrites. When a nerve impulse reaches this region, it causes the release of these neurotransmitters into the space between the dendrites and dendrites of other neurons. This space is known as a synapse.

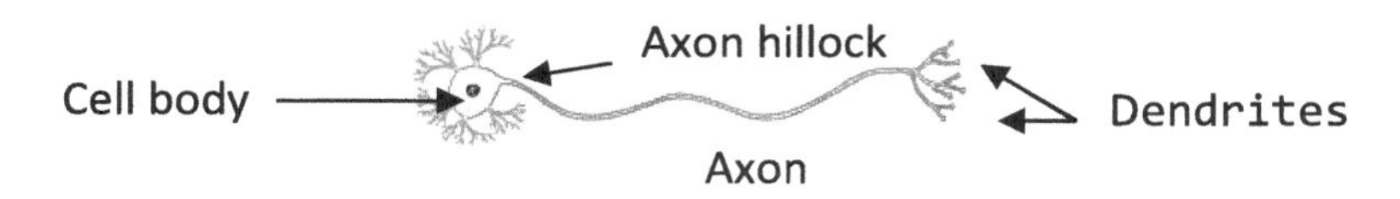

Figure 1. A typical nerve cell in the brain known as an interneuron

5 Neural Pathways

Neurons follow an all-or-none law in the production of a new impulse at the axon hillock. If enough neurotransmitters cross the synapse within a short span of time, usually in milliseconds, they will cross a critical threshold that will fire off an impulse. The application of this all-or-none law in producing a new impulse is what makes the neuron act like an off-on switch. Neurocomputational scientists have used this idea in simulating the process with computers (Hawkins, 2004). They have found that only 3 layers of "neurons" are needed to produce a working model of a neural circuit that can carry out pattern recognition and assign a name to an object defined with certain characteristic properties to a particular "neuron".

We can consider learning as the formation of new neural pathways along various sequential combinations of nerve cell bodies firing off electrical impulses along their axons to stimulate release of chemical neurotransmitters at the gaps or junctions between the nerve cells known as synapses.

The total effect of all the neurotransmitters reaching the next neuron within a short span of time will determine whether the signal is passed on to this next neuron. If the amount of depolarisation of the nerve cell membrane reaches a critical level or even beyond this threshold, it will cause this next neuron in line to fire off an impulse along its own axon.

6 Wiring and Firing Together

This has led to a common saying in neuroscience that nerves that wire together will fire together. What is meant is that the nerves that are part of a particular neural pathway will strengthen their ability to respond more strongly and with greater frequency and regularity than ever before as a single unit in response, the more often that pathway is used. This is the neurocognitive understanding of drill and practice that is carried out as part of a person's schooling.

Each fact or item of information, which includes the individual letters of the alphabet and numbers as well as mathematical symbols like those for addition, subtraction and decimal points is captured as a specific sequential pathway of nerves. Other pathways of nerves are connected to these lower level or foundational order of circuits or pathways in their own hierarchies giving rise to higher and higher orders of thinking and conceptualisation, forming what we commonly refer to executive functions of the brain or higher order thinking skills.

However, it is essential that the practice is at a level deep enough for further learning and not just carried out inefficiently, such as with rote memorisation. This deep learning or deep practice has to be customised for individual learners as each person will vary in terms of pacing and types of difficulty with problems in understanding.

7 The Myelin Sheath

The myelin sheaths that surround the nerve axons grow by winding around the axons and thus forming a sheath that spirals around an axon and envelopes it in numerous layers, like the skin of an onion. The myelin sheaths around the axons of nerves that form pathways that are seldom used have only a few layers (less than ten), whereas the myelin sheaths around the axons of nerves that form pathways that are very often used can have more than fifty layers around them. The greater the number of layers, the greater is the intensity and speed of transmission of the nerve impulses along those pathways.

8 New and Emerging Ideas in Neurocognition

The human brain is distinctively different from the brains of other species in the animal kingdom particularly with regard to the size of its two cerebral hemispheres. Figure 2 shows the four major lobes that make up the left cerebral hemisphere. Of definite interest to us is the area known as the prefrontal cortex (PFC) where the neural circuits are involved in conscious decision making and higher order thinking skills. In relation to the teaching of mathematics, the neural circuits in the PFC form what is often referred to as working memory or short-term memory. The capacity to hold data is very limited, and especially so if the data or facts appear unrelated to the learner.

The PFC has circuits that are linked to other neural circuits in different parts of the cerebral hemispheres. Circuits in the occipital lobes are involved with processing visual information while those in the temporal lobes process auditory information. The frontal lobe deals with sensory information while the parietal lobe controls motor functions and muscle movements.

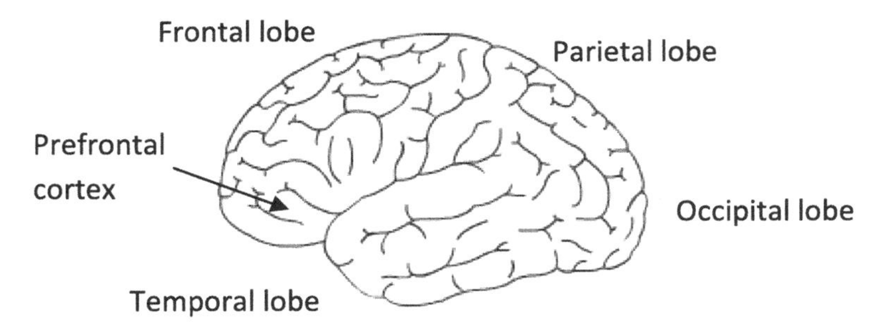

Figure 2. The four major parts or lobes of the brain

Let us take an example of what can take place in our brains when we get up in the morning, and this is illustrated in Figure 3. Neural circuits in the region labelled 1 are active when we wake up on hearing the sound from an alarm that we have set the night before. These sounds are processed in the region numbered 2. When we search for the light switch, the brain circuits at 3 and 4 are active, and when we move around the room to switch on the light, the area numbered 5 in the parietal lobe will control our muscles in the hands and legs. When the light is on, and we pick up our hand phones to check the time and to work out the itinerary for the day, the cognitive functions are being processed by the neural circuits at 6 which is the region of the prefrontal cortex.

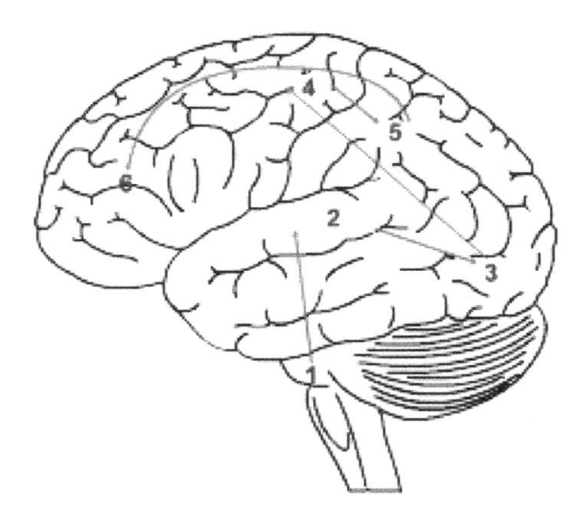

Figure 3. Activation of different neural circuits in the brain on awakening

Appendix A provides further elaboration of the complexity (Spitzer, 1999) to which these neural circuits can give rise to, and it is precisely this level of complexity that enables the human brain to keep on learning well into old age, but at the expense of also having to forget information that is seldom used as the storage for long-term memory has a finite capacity.

9 Deep Practice

Coyle (2009) gives numerous examples of athletes, sportsmen, musicians, chess players and mathematicians who have achieved excellent mastery of their fields of expertise through sustained and focused practice of the complex and difficult aspects of their sports, games or intellectual challenges. The retention of these skills that appear complicated to the public and are viewed as special talents of these masters are attributed to the increase in the myelin sheaths that are formed around the nerves with repeated use of the relevant neural pathways.

10 Neuronal Networks

The nerves or interneurons in our brains are connected to each other to form neuronal networks. The overall physical structure of these networks do not change substantially with age or learning, but these nerve cells are alive, and there are immense microscopic changes at the terminal ends of the nerves which respond to sensory input to the brain. These changes are functional in nature and are essentially changing the wiring of the brain. In essence, each time we learn something new, the brain is being rewired, not physically, but in terms of the pathways that the neural impulses take from one part of the brain to another.

A useful analogy is to think of trains travelling along railway lines, which are fixed in terms of their physical location. However there are switching points along the train tracks which can allow a train to move from one track to another. In this analogy, the path of a particular train from its origin to its destination represents a neural pathway or learning path. The train tracks represent the physical location and structure of the nerves in the brain, while the cell body of a neuron represents a set of switching points along the train tracks.

11 An Analogy of Learning Paths as Learning New Routes of Travel

As we are considering the idea of learning as developing new neural pathways, another appropriate analogy in the real world would be that of learning new routes to travel in a city. This analogy is further developed as follows.

Take the example of Anna, a teacher who has been invited to deliver a keynote lecture in a new city that she has never visited before. On her first morning after arrival, the conference organizers have arranged from a limousine to take her from her hotel to the university where the conference is being held. In this instance, the only thing Anna has to ask the driver is to check whether she is being driven to the correct venue for her keynote speech.

On the morning of her second day, she decides to travel by train as there is a subway station directly opposite her hotel. Along the way, she has to change once to get onto another line. The subway station at her destination is fortunately at the end of this line, where a shuttle bus brings her directly to the lecture theatre at the university for the day's proceedings. This time, the travelling that Anna has to do requires more learning on her part which translates from the different decision steps and routes to take to more neural pathways to be formed for this travel by train.

On the third day, she decides to travel by bus as there is a convenient bus stop just on the road to the right of her hotel. The journey takes longer but is more convenient as she does not have to change buses. The cognitive process however is more difficult as Anna will have to recognise where to get off the bus as the university is not its final stop.

Let us take a final example of Anna deciding to rent a car to drive to the university on the final day of the conference as she is leaving for another city at the end of the morning's proceedings. The neurocognition that she will have to process will be the most difficult of all the various modes of travel over the past few days as she will not only have to learn another different set of routes but also the mental processes needed for driving and observation of the road conditions. This however is the method by which she will really be able to say that she can drive herself around the city without dependence on public transport or the help of others, at least in getting from the hotel to the university.

We have thus found different levels or gradients of difficulty in learning using this example, namely, from being chauffeured around to travelling on routes via train and bus and autonomously by driving one's own car. Being chauffeured around represents a coaching environment where the learning is determined mainly by rote learning, usually with little understanding. This may be necessary in certain circumstances when time is very limited or concepts are very complex, and a formulaic approach or learning by rote memorisation may be a way to deal with the situation.

The mode of travel by train or bus requires much more understanding and this represents a balanced environment where there is both instruction to guide the students in their approach to learning a subject with regard to its fundamentals, complexity and limits, together with constructive elements where the students have time to work things out for themselves. The travel by car or even when one has to walk from one place to another represents an autonomous environment for learning where the onus and responsibility is greatest on the learners to observe and think and work things out for themselves. This also represents an environment where there is maximal possibility for creativity and understanding of complex concepts and principles.

The example of Anna describes 3 types of learning environments in terms of the students being "Coached", being "Balanced" and being "Autonomous". At one end, the simplest environment is represented as a level which can be thought of as that of being cared for by a "coach". The balanced environment has a mixture of instruction and construction in terms of guidance from teachers and freedom to experiment for the learners. The environment that allows for greatest autonomy is also one that is time consuming, but which has the greatest potential for developing creativity and understanding of complex situations.

12 An Example of Collaborative Learning

We can pose a problem familiar to many teachers of mathematics like the following.

> *A grandmother is celebrating her 86th birthday with four of her grandchildren who are 19, 20, 20 and 21 years old. How old will she be when the sum of all the ages of these 4 grandchildren will equal hers at that time?*

The solution to this problem which should seem quite evident to most mathematics teachers as being 88 years is quite vexing when posed to the general public.

However the following example is actually quite easily acquired if the members of the public had spent some time experiencing the problem and experimenting with solutions in their earlier school years or even in their spare time.

> *Let's follow Anna, the teacher, in her home city now as one evening, she sits around a table at the cafeteria by the university sports field with students Ben, Chloe, Din and Emma after a game of frisbee which Anna participates as a player and a coach on the varsity team. Anna tells the students that she has to leave soon for a family dinner as they are celebrating her grandmother's 86th birthday.*

> *Chloe notes that in 2 years' time, the combined ages of the 4 students will equal the age of Anna's grandmother as Emma is 19, and Din is 21 now while she and Ben are both 20 years old. Emma as usual vocalizes her thinking by confirming that their total is now 80 years, and that the total will be 84 next year and 88 the year after, thus equalling that of Anna's grandmother two years ahead.*

> *Ben then wonders whether there is a way to work this out irrespective of the starting ages, so he gathers their 5 empty drink*

cups, wipes them dry, and writes numbers on small slips of paper that are placed inside each cup. Din is deep in thought wondering if this mathematical puzzle might have practical strategic applications in the real world like finance and politics, but is game enough to join in the experiment to work out the underlying mathematical principles behind the problem.

Anna already has begun thinking in terms of the age of each person plus x number of years, and placing the sum of the 4 students on one side of the equation to match the age of the age of the grandmother plus x number of years on the other side of the equal sign, but wisely chooses not to remind them of algebra as a possible solution to the puzzle, and cheerfully takes leave of them as they stay on to work and talk it out.

Exploration and experimenting however takes a considerable amount of time if carried out as individual exercises, but with the increasing power of information technology and social networks in aiding collaborative endeavours, the useful sharing property of this resource can be tapped for the good of the students taught by innovative teachers.

As Colvin (2008) has noted, talent is not enough and not even necessary in raising the mathematical ability of the general public. What is required is adequate time for exploring and experimenting with solutions for any particular mathematical problem. The general conceptual solution to such a problem has far more practical applications in reality than is realised by the public as these concepts can be applied to equivalent problems even in fields like politics, health and economics where time is still the relevant factor but the ages of the people can now represent financial resources like cash and factory equipment.

13 Use of Technology

Modern advances in information technology have great implications for teaching and learning in the 21st century. The ability to develop multimedia applications at relatively little cost compared to what was available even 20 years ago by very small teams of programmers or even just individuals means that we have the available means to use technology well to help each student learn at his or her own pace to reach his or her own level of achievement in understanding and applying that knowledge in practice to reach outstanding levels of mastery efficiently and effectively.

Clark, Nguyen, and Sweller (2006) have described numerous techniques to improve the user interface when developing multimedia courseware for learning, in order to compensate for the small capacity of working memory in the prefrontal cortex and therefore reduce the cognitive load that learners usually have to struggle with when first learning new formulae, concepts and ideas.

Medina (2008) has also described various useful rules for using the brain, such as the idea that the neural or learning circuits are wired differently in each person, and what one experiences in real life literally rewires the circuits, and hence experience does make a considerable difference when it comes to learning something new. Other useful suggestions include the rule about repeating things in order to remember better as this will overcome some of the limitations of short-term memory, as well to remember to repeat things within 30 seconds, as it is the typical retention time for holding seven pieces of unrelated information in the average human brain. Ambrose, Bridges, and Lovett (2010) have also contributed useful knowledge about smart ways of teaching with their explanation of how learning works based on seven research principles.

14 Neurocognition, Learning and Mastery

Recent research and finding in the fields of cognitive neuroscience, neuropsychology and neurocomputation have increased our appreciation of the importance of ensuring that pupils understand new meanings and concepts as there is a distinct correlation with the development of new neural pathways in the brain.

It is also important to rehearse and practice working with the new information and knowledge to enable the myelin sheaths that surround the nerve axons to grow so that the nerve impulses along the pathways increase in intensity and speed of transmission. This will enable the new pathways and methods to eventually replace unwanted older habits of thinking.

Encouraging the habitual practice of reflection both enables new neural pathways to be formed and to be strengthened too. These concepts can be applied in various ways in schools so that a quantum leap in the education of each individual pupil to his or her own level of competence in understanding and at his or her own pace will become a reality in the classroom of today.

15 Conclusion

The practice of reflection results in important benefits to any student who practices this regularly, not just in strengthening the new neural networks that are formed every time some new fact, information, skill or process is learnt but also in developing the creativity to think beyond the borders and limits of everyday cognitive routines of habit.

Reflection is the intellectual exploration of the various pathways of thought using our conscious mind, as it involves the prefrontal cortex, and the use of questions is what gives us insight into new concepts, principles and ideas. Properly taught, reflection is a very valuable and increasingly vital skill that our students can develop which will lead to a significant leap in the ability to master mathematical problems that require creativity and insight. This is a new field that will be important in this century of a globalised, collaborative community of nations competing in an increasingly technological world.

Acknowledgements

The author wishes to thank Professors Berinderjeet Kaur and Toh Tin Lam for their invitation to deliver a keynote address, on neurocognition and reflection, at the inaugural AME-SMS Conference 2012, and for the subsequent invitation to compose this chapter for the Yearbook. Thank you also to all the medical students and doctors and teachers who have asked questions on this subject as they have greatly contributed towards clarifying my ideas on how to explain it in terms of their practical applications.

References

Ambrose, A.A., Bridges, M.W., & Lovett, M.C. (2010). *How learning works: 7 research-based principles for smart teaching*. San Francisco: Jossey-Bass.

Clark, R., Nguyen, F., & Sweller, J. (2006). *Efficiency in learning: Evidence-based guidelines to manage cognitive load*. San Francisco: John Wiley & Sons.

Colvin, G. (2008). *Talent is overrated: What really separates world-class performers from everybody else*. New York: Penguin Books.

Coyle, D. (2009). *The talent code: Unlocking the secret of skill in maths, art, music, sport, and just about everything else*. London: Random House Books.

Drake, R.L., Vogl, A.W., & Mitchell, A.W.M. (2010). *Gray's anatomy for students*. Philadelphia: Churchill Livingstone Elsevier.

Hawkins, J. (2004). *On Intelligence*. New York: Times Books, Henry Holt & Co.

Medina, J. (2008). *Brain rules*. Seattle: Pear Press.

Spitzer, M. (1999). *The mind within the net: Models of learning, thinking and acting*. Cambridge, Massachusetts: The MIT Press.

Appendix

The following four diagrams illustrate the concept of how neuronal networks have been modelled in computer simulations that may give us a better idea of the way learning takes place in our minds and brains. All the diagrams show a network of nerves that are connected in a configuration that does not physically change. What changes however are the routes that nerve impulses can travel from one neuron to another.

In Figure 4, a simple pathway is illustrated in which the nerve impulses travel sequentially through three neurons.

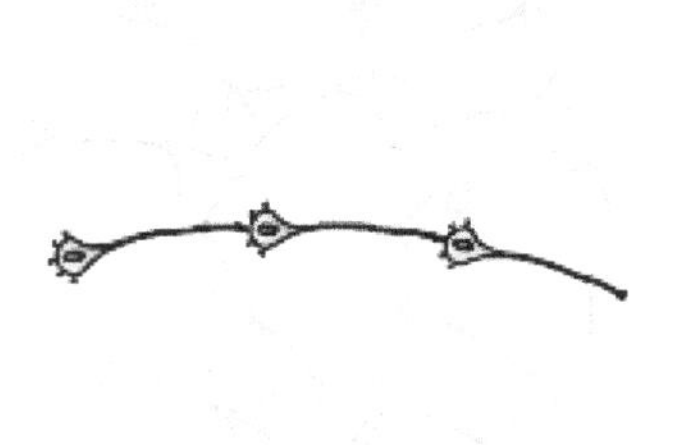

Figure 4. Three neurons forming a sequential pathway for serial processing of information

Figure 5 illustrates the concept of divergence in which the nerve impulses from the first neuron (which is on left of the diagram) transmits the impulse to three other neurons which then send nerve impulses that travel sequentially through other neurons.

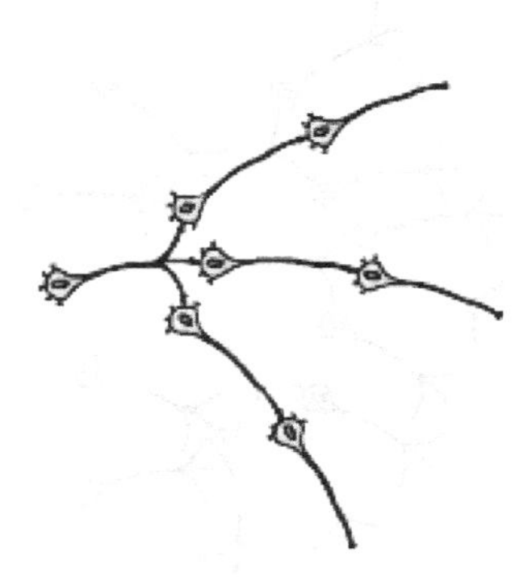

Figure 5. Neurons in the neuronal network showing divergence

Figure 6 shows a pair of examples to illustrate the concept of convergence in which three neurons form neural connections with one neuron. If the nerve impulses from some or all of the neurons arrive the same time, and if their combined strength is above a critical threshold, the one neuron will then transmit an impulse through its axon to other neurons. If the combined strength is below the critical threshold, the one neuron will not send an impulse through its axon. This is an example of the application of the principle of the all or none law for nerve transmission.

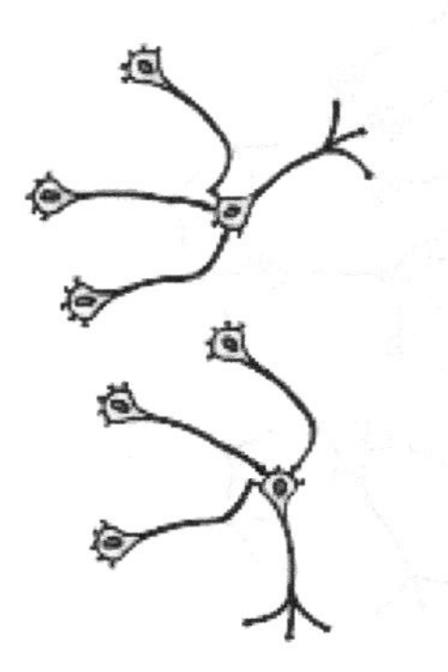

Figure 6. Neurons in the neuronal network showing convergence

Figure 7 is an example of feedback or reverberation, where the second neuron in the has connections that lead back to the first neuron. This feedback may be positive or negative. A negative feedback will reduce the capacity of the first neuron to send more nerve impulses, whereas a positive feedback to the first neuron will increase its tendency to send even more impulses either in terms of frequency or strength of impulse.

Figure 7. Two neurons involved reverberation that enables feedback to the first neuron

There are other patterns of neuronal transmission in such networks, but these few examples should suffice in illustrating how their complexity can result in numerous variations of pathways that can match the vast amount of new information that the brain receives each day. Two other concepts that are difficult to illustrate but which are important and should be kept in mind is that the nerves and neuroglia are living cells and so can grow in response to the impulses that travel along the circuits that they form. The general concept for neurons is that they can increase the amount of terminal connections at the dendrites where there is a corresponding increase in neural transmission along their axons. The general concept about neuroglia is that the myelin sheath that they produce to surround the axons can increase in amount again in response to a corresponding increase in neural transmission along the axons. Coyle (2009) has estimated this increase to about 3000 times and likened it to the equivalent of a broadband connection, as there is both an increase in the speed of transmission of nerve impulses as well as a decrease in the refractory period (the waiting time between successive impulses).

Chapter 3

Working with the Whole Psyche: Nurturing Reflective Learners

John MASON

The whole person is an intricate interweaving of intellect-cognition, emotion-affect and behaviour-enaction with attention and will. By working on some mathematics together and reflecting on our actions, ways of mathematically working with the whole student, such as directing attention, promoting sense-making, provoking engagement, developing dispositions, nurturing reflection and achieving success are likely to come to the surface.

1 Introduction

In 1945 George Pólya published his little book "How To Solve It" in which he resuscitated the word heuristics and suggested four phases of problem solving: Recognise the problem; Devise a plan; Carry out the plan; Look back over the work. In his subsequent books (Pólya, 1954 vols. 1 and 2; 1962), he provided more and more examples of specific strategies, applications of heuristics and mathematical themes. His books contained a reflective element in the form of his commentaries on the problems, but did not directly prompt the reader to reflect except by imitation.

Having been introduced to Pólya's ideas through his film Let Us Teach Guessing (Pólya, 1965), I found myself at the Open University when the first mathematics course was being assembled, and was asked to design a one week summer-school for some 7000 students spaced over 11 weeks at 3 different sites. The programme I devised began with the film and then attempted to use a similar approach to consolidating and explicating mathematical topics arising in the course. It took me some years to realise that many of the tutors we hired, while experienced teachers with a strong mathematical background, nevertheless did not naturally draw student attention to the processes highlighted by Pólya, and which, in later years, I have come to see as natural powers that can be honed and developed as part of learning to think mathematically.

Ten years later, while developing a course for teachers and would-be teachers called Developing Mathematical Thinking (Floyd, Burton, James, & Mason, 1981), we decided to include in each week's work an evening's work on a mathematical problem. In order to choose which problems to present, and what commentary to offer about what affordances were available, Leone Burton and I, and subsequently with Kaye Stacey, decided to assemble a text from which we could then choose what elements to put in the course. The result was Thinking Mathematically (Mason, Burton, & Stacey, 1982/2010), recently extended by an extra 70 new problems, and with a guide to the relevance of each of the problems to readers with different curricular and teaching interests. There is also a new preface, and a link between the discourse of mathematical processes and that of mathematical power and themes. In that book we explicitly develop an experiential stance, inviting the reader to engage with problems and then to take a reflexive stance, exemplified by our commentaries on what we have seen people do with those tasks. The commentaries really only make sense if the reader has themselves worked on the problems. The same is true of this paper.

The theme of this chapter is, as the subtitle says, nurturing reflective learners. The message is brief, but the application is subtle, calling as it does on the teacher's reflexivity concerning their own thinking so as to sensitise themself to what learners may be experiencing. Message: One thing we do not learn from experience is that we do not often learn from experience alone. A succession of experiences does not in itself add up to an experience of that succession. Something else is required. In school, students are asked to undertake tasks; engagement with tasks leads to activity; activity provides experience; but in order to learn from that experience, it is at least helpful and often necessary to withdraw from the activity and to adopt a reflexive stance, asking oneself questions such as

- What was ineffective and what was effective?
- What got me stuck, and what got me unstuck?
- What powers did I use?
- What themes did I encounter?
- What would I like to have come-to-mind in the future?
- What did I learn about myself (dispositions, propensities, habits, etc ...)?

Most importantly, a few seconds spent imagining myself acting in a more informed manner in the future can be worth hours of wasted effort failing to learn from experience in the past. It is not the task that is rich but the way in which the task is used.

2 Approach

My approach is experiential. In what follows I shall offer some tasks to be worked on, which from experience I know have the possibility of affording access to important aspects of mathematical thinking. I will offer some commentary but this will only make sense to those who have themselves engaged with the tasks. Because of the necessary brevity of a chapter it will not be possible to do more than touch on two or three aspects of mathematical thinking. The remarks here can be taken as a taster of a general way of working.

The theoretical basis for my approach has a long and ancient history, but there is no time to talk about that. Suffice it to say that when a teacher works with students over a period of time, there can be a process of integration through subordination as Gattegno (1970) put it, invoking what Wood, Bruner, and Ross (1976) called scaffolding (teacher prompts intended to bring to mind what students can already do but need to be reminded of) augmented by the equally vital process of fading (using less and less direct prompts so as to transfer the initiative to students). This is what van der Veer and Valsiner (1991) propose is the core meaning of 'working in the Zone of Proximal Development' as intended by Vygotsky (1978, pp. 84-88; see also Mason, Drury, & Bills, 2007). My colleagues and I at the Open University formulated this independently for ourselves and for the teachers we worked with in the three-term framework directed–prompted–spontaneous (Love & Mason, 1992). By this we meant using direct questions and suggestions at first, which over time become increasingly indirect ("what question am I going to ask you?", "what did you do yesterday when this happened?" etc.) until students are using them spontaneously for themselves. For an excellent example regarding the prompt "What is the same and what different about …", see Brown and Coles (2000).

3 Preliminary Tasks

All of the following tasks have been used in some form with students, but they are formulated here for secondary teachers and would have to be modified and augmented to be appropriate to any specific group of students.

3.1 *Arithmetical relations & properties*

> What is 53 + 74 – 29 – 53 + 29?
> Is it true or false that 137 + 481 = 481 + 137?
> Is it true or false that 37 + 94 = 34 + 97?

These probes are representative of a class of research tasks used by various people (Molina 2007; Molina, Castro, & Mason 2007, see also Mason, Stephens, & Watson, 2009). When calculations like the first are offered to students, some embark on calculation, others start calculating and then stop, and others 'see' what is going on and simply give the answer. Their actions reveal something of their disposition (for example to jump in and do the first thing that comes to mind) and also something about their awareness of arithmetical relations. The second one causes no difficulty for teachers, but that is because they have learned to 'read the question' before embarking on an action. The third one requires some movement of attention: discernment of detail, recognition of relationships, perception of properties as being instantiated and awareness of what place value is about in order to avoid simply doing the computation.

The deeper or perhaps inner task (Tahta, 1981) is to articulate for each equation a generality of which the equation is a particular instance. Arithmetic is for me the study of actions on numbers, as a way to study the properties of numbers. Calculations are a by-product. Mastering arithmetical procedures is valuable so as to free attention to be awake to recognising relationships and to perceiving properties as being instantiated. This is precisely what is required for the third calculation.

I use 'parking' as a label for the flexibility to be able to delay carrying out the first action that comes to mind in favour of asking yourself whether there might be a better, more effective, more efficient approach. It is readily distinguished from the state of being stuck when no action comes to mind.

Reflecting. Catching a moment when I am about to embark on a calculation or other action, and bringing to the surface a question about whether there might be a better alternative (parking) is fodder for reflection. The most important aspect of reflection is to then imagine myself in some similar situation in the future, enacting 'parking' or whatever action was appropriate.

Another important role of reflection is placing what I have done in a more general context. Can I construct a similar example for myself? Can I construct an example which is 'peculiar' in some way (Bills, 1996)? Can I express a generality? As with any exercise, the answer itself is not as important as being aware of a class of tasks of which it is representative. The student who goes into an examination and is surprised by some or all of the questions is in a weak position; the student who recognises the type of the question and can say to themselves "I could construct a much harder example than this ... shall I be gracious and do it for the examiner?" understands, indeed 'stands over' or 'is on top of' their studies. They are likely to do well.

3.2 *Recognition1*[1]

> Given that 257257 is divisible by 7, what is the remainder on dividing 25 725 756 by 7?

Even if you distrusted the initial assertion and tested it for yourself, there is no need to start again with the division. But in order to avoid a calculation you need to discern the two numbers, and recognise a relationship between them: seeing the second number as of the form 100x first number + 56, so it too must be divisible by 7.

The astute reader notices that the first number has three repeated digits and so must be divisible by 1001, and a worthwhile fact to know is that 1001 is 7 x 11 x 13. So for example, the remainder on dividing the second number by 11 will be 1 and by 13 the remainder will be 4.

This task underlines the point that when doing computation, it is vital to be on the lookout to recognise relationships and to see these as instantiations of properties. It is through reflection in the present on actions undertaken recently that improved actions may come to mind in the future (Mason, 2002). Schön (1983) distinguished between reflection on action, retrospectively, and reflection in action which to me refers to being awake to the possibility of choosing to act, in-the-moment, rather than working on automatic pilot under the direction of habit.

Again, the teacher who is promoting mathematical thinking will want to ask students to create their own task like this. Indeed, having used this as a pedagogic strategy as part of explicit reflection, and then making the prompts gradually less and less explicit, students can take over the initiative and make example-creation part of their natural study technique.

[1] I got the idea for this task from my wife, Anne Watson.

4 Interlude on the Structure of the Psyche

There is a stanza from the Rg Veda (Bennett, 1964) which speaks directly to the enterprise of nurturing reflective learners.

Two birds, close-yoked companions,

Both clasp the self-same tree;

One eats of the sweet fruit,

The other looks on, without eating.

There are of course several ways to interpret this, but one way is to see the bird eating the sweet fruit as the part of me, the self, that is engaged in activity. The other bird is an inner witness or monitor whose presence guards against being so totally absorbed that I become lost in the activity. The aim of education is perhaps most essentially to promote and provoke students to develop their inner witness, a second bird which sits 'on the shoulder' and asks questions from time to time, like "why are we doing this?" and "Is this the best action to be undertaking?".

The traditional division of the psyche into cognition, affect and enaction, or as I prefer, awareness, emotion and behaviour, has ancient roots. For example in the Upanishads (Rhadakrishnan, 1953, p. 623) the psyche is likened to a chariot:

The chariot itself can be seen as the body, and by extension, behaviour. The horses are often seen as the senses, and since they are what pull the chariot along, by extension, the emotions which is the source of energy which our various selves call upon. The shafts are the automated habits by means of which we engage in activity. The driver is intellect or awareness. The reins refer to mental imagery as the means by which the driver influences and directs the horses. The owner is the will, manifested through attention, for as William James (1890) noted, "we are our attention; we are where our attention is".

My suggestion is that where a teacher is able to engage all the aspects of the students' psyche, engagement and significant learning are likely to take place. Furthermore, training students to be dependent on the textbook or the teacher at best trains behaviour but does not promote creativity. Provoking students to take initiative, to learn how to learn mathematics, to make substantial and significant mathematical choices contributes to the development of the full person rather than simply trying to get them over the next hurdle.

5 Mathematical Themes

One of the most pervasive mathematical themes is that of 'doing and undoing'. Whenever you have an action that achieves a result, it almost always proves fruitful to ask yourself whether, given the result you can find all the 'givens' that lead to that result. Doing and Undoing as a theme starts very young.

5.1 *Doing & undoing additively*

What action undoes the action 'adding 3'?
What action undoes the action 'subtracting 5'?
What action undoes the action of 'adding three then subtracting 5'? (express it in two ways)

This is not a challenging task for secondary teachers or students. However it is important to recognise that it can be constructed by adding 5 and then subtracting 3' as well as the combined action of 'adding 2'. There is also some creative work to be done in asking what are all the ways in which a pair of actions have the combined result of 'adding 2'.

5.2 *Doing & undoing unexpectedly*

What action undoes the action 'subtracting from 7'?

This certainly caught me out the first time I heard it. The first response by many people arises from the running of habit based on familiarity with language: "adding to 7". A more careful interpretation leads to recognition that this is an action that undoes itself, and could lead to a search for other more complicated actions that similarly undo themselves.

5.3 *Doing & undoing multiplicatively*

What action undoes the action 'multiplying by 3'?
What action undoes the action 'dividing by 5'?
How else can the action 'multiply by 3 then divide by 5' be described?
What action undoes the action of 'multiplying by 3 then dividing by 5'? (express it in two ways)
What action undoes the action of 'multiplying by 3/5'

Here the careful undoing through reversing the actions provides an algorithm, while the succinct version reveals that the action is 'dividing by 3/5'. This is one way to justify the playground rule 'flip and multiply' for dividing by fractions.

If you are content to slide onto the next task, then you have missed an opportunity! What is the analogue in multiplication for 'subtracting from 7'? In English it is particularly confusing because the phrase 'dividing into 24' is ambiguous: does it mean divide into 24 equal parts, or does it mean divide the number into 24?

Why is there no analogue to 'subtracting from 7' and 'dividing 24 by' for exponentiation?

5.4 *Reflections*

There have been several opportunities to catch yourself making conjectures, perhaps even 'parking', and recognising personal dispositions come to the surface, manifested for example in terms of like and dislike of my tasks. In lectures and workshops I use directed-prompted-spontaneous to remind myself to ask "what question am I going to ask you?" as a reminder to the audience to pause for a moment and to adopt a reflexive stance.

6 Geometry as Context

Geometrical reasoning is just as fecund with opportunities to learn from experience as arithmetic or algebra because as a domain it is concerned with the multitude of geometric relationships and properties.

6.1 *Alternating sums of squares*

Imagine a triangle. Let P be the intersection of the perpendicular bisectors of the sides (the circumcentre). It is obvious that if squares are erected on each of the segments comprising the edges of the triangle, then the sum of the areas of the squares in cyclic order around the triangle, with alternating + and – signs, must be zero.

Now let P be any point in the interior of the triangle. Drop perpendiculars to the edges, erect squares on each of the segments formed, and form the alternating sum of areas. What is the result? Why?

No diagram is provided because imagining the figure strengthens your power to form mental images (not necessarily pictures as they may have verbal and kinaesthetic components).

To make progress it advisable to draw in certain objects that then reveal relationships. The flexibility to look at the same thing in two different ways (here, arising from the squares with positive sign, and the squares with negative sign) is a desirable trait and constitutes a useful mathematical heuristic, particularly when counting things.

But there is more.

6.2 *More alternating sums of squares*

> Can your result be extended to pentagons, and beyond?
> Does it matter whether the point is in the interior of the figure?
> What happens if the point P coincides with a vertex?
>
> If the edges of a triangle have been divided into pairs of segments so that the alternating sum around the triangle of the squares on those segments is zero, must the perpendiculars to the edges at the division points coincide in a single point?
>
> Show that a quadrilateral for which the alternating sum of squares of the edges is zero if and only if the diagonals of the quadrilateral are perpendicular.
>
> Show that any even sided polygon with a zero alternating sum of squares can be decomposed into a family of quadrilaterals with that property glued together along common edges.

All of a sudden there is an explosion of investigative possibilities. I like to (mis)quote a line from the English poet John Dunne to the effect that "No task is an island, complete unto itself". Any task with pedagogic potential lies in a domain of multiple variations and extensions. Here there are opportunities to work on characterising and extending.

6.3 *The carpet theorem*

Everyone knows what happens when you put down carpets and then move them so that they overlap a bit. But turning it into a significant mathematical theorem might be a surprise.

Two areas

Which is greater, the darkly shaded region or the lightly shaded region in the first diagram?

What is the relationship between the areas of the darkly shaded to the lightly shaded regions in the second diagram?

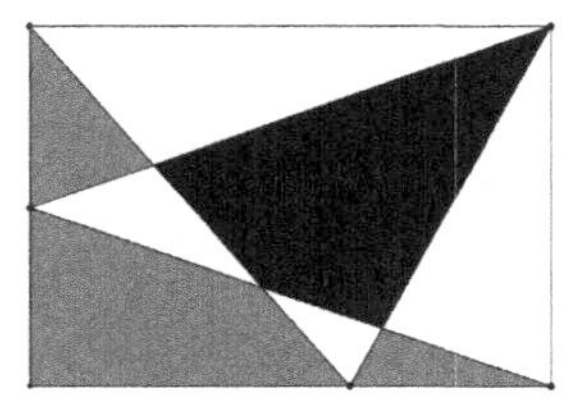

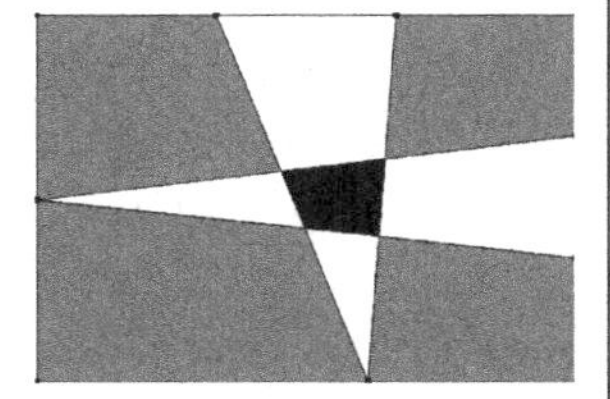

The answer to the first is not immediately apparent from direct perception, though the astute reader or student will second-guess that they must be equal or why would anyone ask. That is why the second one is presented alongside. There the bases of the triangles are each one-third of the edge of the rectangle, but there is an implied generality to make them specified fractions of their respective edges.

The carpet theorems

In a rectangular room there are two carpets that together exactly cover the floor without overlap. One carpet is moved so that it partly overlaps the other. How much floor has been revealed?

Imagine you have two carpets in a room, possibly overlapping a bit, possibly not. You adjust the carpets so as to change the amount of overlap. By how much as the area of the floor not covered by the two carpets changed?

Now apply these to the Two Areas tasks.

In the first figure, it helps to discern two white triangles (as carpets) which have equal area and together exactly cover the floor. Adjusting one carpet (preserving its area) leads to an overlap region (darkly shaded) and part of the floor revealed (lightly shaded). In the second figure, seeing the lightly shaded region as the exposed floor, the two white carpets each account for one sixth of the whole, so the overlap is the same as the amount of floor beyond five sixths that is exposed. Thus the lightly shaded region is five sixths more than the darkly shaded region.

7 Area and Perimeter as Context

It is well known that some students confuse area and perimeter. One reason may be that the two topics are often taught together; another could be that students react to triggers in problems without allowing themselves time to gaze and to think about what they want to know, and what they already know (Mason, Burton, & Stacey, 1982/2010). The following task has a generic structure derived from Tirosh and Tsamir (personal communication).

7.1 *More or less (perimeter and area)*

Shade in shapes in the chart that have the corresponding relationships of area and perimeter.

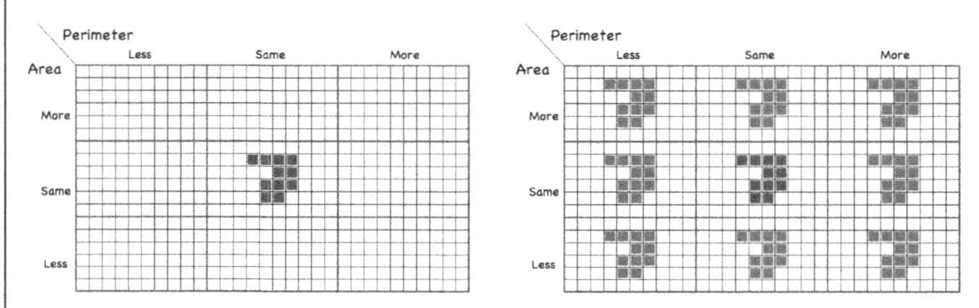

In each case, how few changes can you make from the original (shown in the right hand figure for ease of reference)?

Can you always fill out the chart no matter what rectilinear shape you start with in the centre?

The applet (Mason, 2012a) provides the starting shapes in each cell so that you can keep track of how many changes you make. You can also change the starting shape in the centre.

Reflection. What actions did you use to reduce or increase perimeter while maintaining area? What actions reduce or increase area while maintaining perimeter? How do these relate to actions which increase or decrease both, or which increase one while decreasing the other? It is in becoming aware of the actions that were effective that real learning of substantial mathematics takes place. This is a form of reflective abstraction as described by Piaget (2001; see also Dubinsky 1991a, 1991b; Simon, Tzur, Heinz, & Kinzel, 2004).

The More-or-Less structure can be used for other topics. For example, given a pair of numbers in the centre such as 12 and 4, let the columns refer to smaller, same or larger difference and rows correspond to smaller same or larger ratio. How little can be changed and in what different ways for each cell? When is it possible to preserve one of the numbers? Can the pairs in each cell be integers?

8 Recognising Types of Numbers as Context

When arithmetic is seen as the study of properties of numbers and properties of actions on numbers, multiple opportunities arise for practising computations while developing expertise in conjecturing and expressing generality.

8.1 *Four consecutive sums*

> What numbers can be expressed as the sum of four consecutive numbers?
> Express 524 as a sum of four consecutive numbers.

The first part can be approached empirically, asking yourself what is the same and what is different about numerous examples. More efficiently it can be tackled through tracking arithmetic, that is by using a base number but isolating it from subsequent calculations, leading to expressing a generality through seeing the 'base' number as an aspect that can be varied but leaving a property invariant. Seeking what is invariant and what can change (invariance in the midst of change) is a pervasive and ubiquitous theme in mathematics.

A study involving large numbers of secondary students in the UK revealed that although many reached a conclusion that it was numbers that are two more than a multiple of 4 that can be the sum of four consecutive numbers, very few thought to use this to address the question of whether a specific number (44, or here, 524) could be so expressed

(Hoyles & Healy, 1999). Most tried to find four numbers, rather than applying their recently expressed generality (i.e. starting with 44/4 and trying to build around that).

Learning to park the first idea that comes-to-mind is a really useful disposition to develop! Although not all even numbers can be expressed as the sum of four consecutive, perhaps there other numbers m for which the sum of m consecutive numbers is divisible by m, leading to characterising those numbers.

8.2 *Consecutive sums*

> For what numbers n does n divide the sum of any n consecutive numbers?
> In how many ways can a number be expressed as the sum of consecutive positive numbers?

This is written up in Mason, Burton, and Stacey (1982/2010). The strategy of being systematic is beneficial, but so is the strategy of 'branching out' and trying something bigger. Again it is not the doing of examples per se but what the student does during and with the examples. The purpose of specialising is to detect and express an encompassing generality. It is not simply a matter of getting some answers and then trying to spot a pattern in the answers (what Hewitt (1992) calls 'train spotting') but rather of 'Watch[ing] What You Do' (Mason, Graham, & Johnston-Wilder, 2005) as you do an example, to try to detect structural relationships. Often the body acts in ways that if attended to, display an awareness of generality that cognition may not yet have caught up with. This matches what Courant (1981) said was Hilbert's explicit method: to select an example that was not too simple, and to understand that thoroughly.

Meta-comment. My approach involves identifying specific actions, powers, or themes that contribute to and enrich mathematical thinking, and then look out for tasks in which most people naturally engage that action, use that power or encounter that theme. On reflection I then draw their attention to what they did spontaneously, and suggest that this might be useful in the future. To emphasise this, I can even invite them to imagine themselves stuck on a problem in the future, and to imagine themselves having come-to-mind to try a simpler example with view to re-generalising for themselves. In the near future I can use an indirect prompt ("what did you do last week …?") as a reminder, and as the prompts become less and less direct, I can expect students to have internalised the advice. Occasionally I may need to prompt recall of that advice, but as indirectly as possible so as to underpin their internalisation.

8.3 *One more than the product of four consecutive numbers*

> What numbers can be expressed as one more than the product of four consecutive whole numbers?

If you immediately thought of writing down an algebraic expression, then parking that idea and asking yourself what sort of a property you might be seeking might first lead you to some careful specialising, so that you have a sense of what a reasonable conjecture might look like.

If you are confident with symbols and proceeded algebraically, then you probably wrote down a general expression and then gazed at it for a while. What usually comes to mind then is to expand the brackets and create a polynomial. Most people who do this are none the wiser: you don't know what properties to expect, and you don't have relationships to recognise which might inform that. Alternatively, anticipating an ugly quartic might lead you to specialise immediately but with an algebraic perspective on the particular cases.

Some people start with some specific examples (specialising or particularising). Those with experience of 'consecutives' might begin with $1 + 0 \times 1 \times 2 \times 3$, $1 + 1 \times 2 \times 3 \times 4$, etc.. The purpose of constructing particular examples is not simply to generate examples or to engage in busy-work, but to try to see what is being exemplified, that is, what structural relationships might be determining the behaviour. Articulating these relationships as general properties provides conjectures to then try to justify logically.

Sophisticated specialisers, having been systematic, try an outlier to see if the relationships they have recognised are likely to hold for all examples. For example, 1 + 9 x 10 x 11 x 12 = 11881 is relatively easy to compute mentally (calculators can be of benefit to generate lots of data, but will not in themselves reveal structural relationships: that requires insight). But unless you have a conjecture from small identifiable cases, you are unlikely to see this as 1092. The real question is how the 109 is related to the original four consecutive numbers (the underlying structural relationship).

8.4 *Sundaram's grid*

The diagram shows the bottom left corner of an infinite grid in which each row and each column are arithmetic progressions.

It is claimed (Ramaswami Aiyar, 1934; Honsberger, 1970; Havil, 2009) that a number appears in the grid if and only if doubling it and adding one is a composite number (not prime). See notes accompanying Mason (2010) for related tasks.

8	25	42	59	76	93	110	127
7	22	37	52	67	82	97	112
6	19	32	45	58	71	84	97
5	16	27	38	49	60	71	82
4	13	22	31	40	49	58	67
3	10	17	24	31	38	45	52
2	7	12	17	22	27	32	37
1	4	7	10	13	16	19	22
	1	2	3	4	5	6	7

Reflection. Can you see Sundaram's grid as an instance or example of doing and undoing? Can you see it as an illustration of the theme of invariance in the midst of change? It is certainly an example of a challenge to detect and express a generality, since the only sensible way to justify a statement about all numbers in an effectively infinite grid is to work with an expression that (re)presents all such entries. Does the conjecture apply to the grid when extended to the left and-or downwards?

8.5 *Generalising patterns from 2*

It seems that

$$2+2=2\times 2$$

$$3+1\frac{1}{2}=3\times 1\frac{1}{2}$$

$$4+1\frac{1}{3}=4\times 1\frac{1}{3}$$

What would the 17th equation look like, and would it be correct? Generalise.

How does the first line fit with your generalisation?

What if the line started with 2/3 or 7/5 or $\sqrt{3}$? What about starting with –1? Or other negative numbers?

It is a shame to present this task in print, because much of what it affords is only available through the way it is presented: part by part; inviting prediction (conjectures); focusing on sense-make (going across the grain not just with the grain as in Watson (2000)); recognising relationships and expressing these as properties so as to instantiate them in other cases; inviting expression of generality (the root of and route to algebraic thinking).

For variations see the applet called "Patterns from 2" (Mason, 2012b). It is perfectly natural for people to spot the relationship and be able to express different equations as conjectures (going with the grain), even where their arithmetic is not up to validating the equation (going across the grain). This affords opportunities to draw attention to the natural power to imagine and to express generality, and to the status of conjectures: following a perceived pattern as compared to knowing a particular equation must be true and knowing that all such equations must be true.

These tasks were offered in order to provide repeated exposure to notions such as 'parking', to the use of natural powers used mathematically such as specialising and generalising, conjecturing and convincing, and to the role of mathematical themes such as doing and undoing, and invariance in the midst of change.

To understand arithmetic requires algebraic thinking (Hewitt, 1998). Stressing that arithmetic is the study of actions on numbers (this is how one studies numbers themselves, and is quite distinct from mastering arithmetic calculations) involves, among other things, drawing attention out of a disposition to calculate and into a disposition to be aware of relationships as properties.

9 Reflection on Nurturing Reflection

To nurture reflection in others can only be done effectively by nurturing reflection in oneself, so that a suitable reflective device comes-to-mind at an appropriate moment, provoking withdrawal from the action in order to become cognisant of the action and its effectiveness in the particular situation. Pausing and imagining oneself engaged in the future, and having that same action come-to-mind makes use of the fundamental human power of mental imagery, which is the principal means we have for planning, for preparing for the future. One way to express this using the etymological root paring is to think of pre-paring for the future by post-paring after the event, so as to increase the chances of paring in the moment. A similar construction can be made with the word re-flecting (see Mason, 2002).

References

Bennett, J. (1964). *Energies: material, vital, cosmic*. London: Coombe Springs Press.

Bills, L. (1996). The use of examples in the teaching and learning of mathematics. In L. Puig & A. Gutierrez (Eds.), *Proceedings of the 20th Conference of the International Group for the Psychology of Mathematics Education* (pp. 2.81-2.88). Valencia: Universitat de València.

Brown, L., & Coles, A. (2000). Same/different: a 'natural' way of learning mathematics. In T. Nakahara & M. Koyama (Eds.), *Proceedings of the 24th Conference of the International Group for the Psychology of Mathematics Education* (pp. 2-153-2-160). Hiroshima, Japan.

Courant, R. (1981). Reminiscences from Hilbert's Gottingen, *Mathematical Intelligencer, 3*(4), 154-164.

Dubinsky, E. (1991a). Constructive aspects of reflective abstraction in advanced mathematics. In L. Steffe (Ed.), *Epistemological foundations of mathematical experience* (pp. 160-187). New York: Springer Verlag.

Dubinsky, E. (1991b). Reflective abstraction in mathematical thinking. In D. Tall (Ed.), *Advanced mathematical thinking* (pp. 95-126). Dordrecht, The Netherlands: Kluwer.

Floyd, A., Burton, L., James, N., & Mason, J. (1981). EM235*: Developing mathematical thinking*. Milton Keynes: Open University.

Gattegno, C. (1970). *What we owe children: The subordination of teaching to learning*. London: Routledge & Kegan Paul.

Havil H. (2009). Sundaram's sieve. *Plus Magazine, 50*(13). Retrieved August 15, 2012 from www.plus.maths.org/content/sundarams-sieve

Honsberger, R. (1970). *Ingenuity in mathematics*. New Mathematical Library #23, p. 75. Washington, DC, USA: Mathematical Association of America.

Hewitt, D. (1992). Train spotters' paradise. *Mathematics Teaching, 140*, 6-8.

Hewitt, D. (1998). Approaching arithmetic algebraically, *Mathematics Teaching, 163*, 19-29.

Hoyles, C., & Healy, L. (1999). Linking informal argumentation with formal proof through computer-integrated teaching experiments. In O. Zaslavsky (Ed.), *Proceedings of the 23rd Conference of the International Group for the Psychology of Mathematics Education* (pp. 105-112). Haifa, Israel.

James, W. (1890 reprinted 1950). *Principles of psychology*, Vol. 1. New York: Dover.

Love, E., & Mason, J. (1992). *Teaching mathematics: Action and awareness*. Milton Keynes: Open University.

Mason, J. (2002). *Researching your own practice: The discipline of noticing*. London: RoutledgeFalmer.

Mason, J. (2010). Available to use online at mcs.open.ac.uk/jhm3/SVGrids/Cindy%20Grids/Sundaram%20Developed/Sundaram%20Full.htmland for downloading (with notes on how to use it) at mcs.open.ac.uk/jhm3/SVGrids/SVGridsMainPage.html

Mason, J. (2012a). More-Or-Less: applet available online at http://mcs.open.ac.uk/jhm3/Applets%20&%20Animations/Applets%20&%20Animations.html

Mason, J. (2012b). Patterns From Two: applet available online at http://mcs.open.ac.uk/jhm3/Applets%20&%20Animations/Applets%20&%20Animations.html

Mason, J., Burton, L., & Stacey, K. (1982/2010). *Thinking mathematically* (Second Extended Edition). Harlow: Prentice Hall (Pearson).

Mason, J., Drury, H., & Bills, E. (2007). Explorations in the zone of proximal awareness. In J. Watson & K. Beswick (Eds.), *Mathematics: Essential Research, Essential Practice: Proceedings of the 30th annual conference of the Mathematics Education Research Group of Australasia* (Vol. 1, pp. 42-58). Australia: MERGA.

Mason, J., Johnston-Wilder, S., & Graham, A. (2005). *Developing Thinking in Algebra.* London: Sage (Paul Chapman).

Mason, J., Stephens, M., & Watson, A. (2009). Appreciating mathematical structure for all. *Mathematics Education Research Journal, 21*(2), 10-32.

Molina, M. (2007). *Desarrollo de Pensiamento Relacional y Comprehensón del Signo Igual por Alumnos de Tercero de Educación Primaria.* Unpublished PhD Thesis, Granada: University of Granada.

Molina, M., Castro, E., & Mason, J. (2007). *Distinguishing approaches to solving true/false number sentences.* Paper presented at CERME 5, Working Group 6, Lanarka, Cyprus.

Piaget, J. (2001). *Studies in reflective abstraction.* Sussex: Psychology Press.

Pólya, G. (1945). *How to solve it: A new aspect of mathematical method.* Princeton: Princeton University Press.

Pólya, G. (1954). *Mathematics and plausible reasoning Vol. 1: Induction and analogy in mathematics.* Princeton: Princeton University Press.

Pólya, G. (1954). *Mathematics and plausible reasoning Vol. 2: Patterns of plausible reasoning.* Princeton: Princeton University Press.

Pólya, G. (1962) *Mathematical discovery: On understanding, learning, and teaching problem solving* (combined edition), Wiley, New York.

Pólya, G. (1965). *Let us teach guessing,* (film). Washington, DC, USA: Mathematical Association of America.

Rhadakrishnan S. (1953). *The principal upanishads.* London: George Allen & Unwin.

Ramaswami Aiyar, V. (1934). Sundaram's sieve for prime numbers. *The Mathematics Student, 2*(2), 73.

Schön, D. (1983). *The reflective practitioner: How professionals think in action.* London: Temple Smith.

Simon, M., Tzur, R., Heinz, K., & Kinzel, M. (2004). Explicating a mechanism for conceptual learning: Elaborating the construct of reflective abstraction. *Journal for Research in Mathematics Education*, *35*, 305-329.

Tahta, D. (1981). Some thoughts arising from the new nicolet films. *Mathematics Teaching*, 94, 25-29. Reprinted in R. Beeney, M. Jarvis, D. Tahta, J. Warwick, & D. White. (Eds.) (1982). *Geometric images* (pp. 117-118). Leapfrogs, Derby: Association of Teachers of Mathematics.

van der Veer, R., & Valsiner, J. (1991). *Understanding Vygotsky*. London: Blackwell.

Vygotsky, L. (1978). *Mind in society: the development of the higher psychological processes*. London: Harvard University Press.

Watson, A. (2000) Going across the grain: mathematical generalisation in a group of low attainers. *Nordisk Matematikk Didaktikk (Nordic Studies in Mathematics Education)*, *8*(1), 7-22.

Wood, D., Bruner, J., & Ross, G. (1976). The role of tutoring in problem solving. *Journal Child Psychology and Psychiatry*, *17*(2), 89-100.

Chapter 4

Knowledge and Beliefs for Nurturing Reflective Learners of Rational Number Concepts

Kim BESWICK

Examples of teaching and learning rational number concepts from Years 2 to 8 (students aged 7 to 14 years) are used as a basis for considering the nature of reflective mathematics learning. It is argued that teachers' knowledge and beliefs are crucial to the development of reflective learners. Work samples and transcripts of learning episodes are used to illustrate behaviours of reflective learners and teacher actions that seem to support reflective learning, and the teacher knowledge and beliefs that appear to underpin these. Teachers need to have a good understanding of the mathematics that they teach and also typical ways that students think about that content, including likely areas of difficulty. They need to be able to analyse students' thinking in the midst of teaching and learning interactions, and respond with appropriate questions, explanations, and representations of the important mathematical ideas. The importance for teachers of reflecting upon their own beliefs about such things as the nature of mathematics and hence what it means to know, learn and teach mathematics, and about the capacities of their students and the importance of them developing a reflective approach to their learning are also considered.

1 Introduction

Reflection has been recognised as crucial for developing deep understanding of the content being studied and of one's own learning. It is important, therefore, that teachers can recognise reflective activity in their students and act in ways that encourage it. Choosing and structuring tasks that provide opportunities for reflection is one aspect of relevant teacher actions and, like all teacher decisions, is underpinned by the knowledge that teachers bring to their work and the beliefs that they hold about their role, their students and how they learn, and the nature of mathematics (Beswick, 2007). Although it has been argued that knowledge and beliefs are equivalent (Beswick, 2011), for clarity they are considered separately here.

2 Teacher Knowledge and Nurturing Reflective Learners

The knowledge needed to teach mathematics has been the subject of much research since Shulman (1987) described seven types of teacher knowledge. These were content knowledge, general pedagogical knowledge, knowledge of students, knowledge of educational contexts, curriculum knowledge, knowledge of the purposes of education, and pedagogical content knowledge (PCK). Mathematics educators have worked to understand how these generic knowledge types apply specifically to mathematics teaching. PCK, described by Shulman as derived from a combination of pedagogical and content knowledge, has been of particular interest to mathematics educators. Ball and colleagues (e.g., Ball, Thames, & Phelps, 2008), for example, have examined the role of mathematical content knowledge in mathematics teaching and elaborated the notion of PCK, conceptualising it as comprising knowledge of mathematics and students, knowledge of mathematics and curriculum, and knowledge of mathematics and teaching. Chick, Pham, and Baker (2006) took a different approach, describing knowledge that is clearly PCK (e.g., describing teaching strategies for particular mathematics concepts or analysing student misconceptions), content knowledge in a pedagogical context (e.g., making connections between concepts), and pedagogical knowledge in a content context (e.g., identifying strategies for engaging students).

All of the aspects of teacher knowledge are needed to nurture reflective learners. Teachers need to know the mathematics that they want their students to reflect on including its connections to other mathematical concepts and its place in the curriculum. They need to know how to engage their students with important mathematics. They must be able to probe students' thinking and have available a range of ways to respond that help students to deepen and extend their understanding. They must know how to explain mathematical ideas in ways that are meaningful for students and have available a range of representations.

3 Teacher Beliefs and Nurturing Reflective Learners

In addition to having the knowledge required to nurture reflective learners, teachers need to believe that doing so is both worthwhile and possible. Although teachers are often urged to adopt particular teaching practices there is evidence that these are effective only if teachers understand the principles (or beliefs) that underpin them (Askew, Brown, Rhodes, Johnson, & Wiliam, 1997; Beswick, 2007; Watson & De Geest, 2005). Hogan, Rahim, Chan, Kwek, and Towndrow (2012) similarly observed that the effectiveness of practices, including those that have routinely been characterised as ineffective, is dependent upon the purposes for which they are used. Teachers' purposes in implementing whatever teaching strategies they choose reflect their beliefs about what is important in mathematics teaching and learning, what they consider to be valuable outcomes of mathematics teaching, and how their students' learn mathematics.

These and all beliefs are not held in isolation but as parts of systems in complex connection with other beliefs (Green, 1971). Beliefs are also highly contextual with different combinations of beliefs being more or less influential in different circumstances. Beswick (2004, 2005) described teachers whose most centrally held beliefs differed according to the class that they considered, with their beliefs about appropriate teaching strategies dependent upon their beliefs about the ability of the class to learn mathematics. Because reflection involves sophisticated thinking that is metacognitive in nature and thus characterised by the ability to monitor and regulate one's own thinking, it may not be considered a realistic aim for young children or for students who struggle to learn mathematics. If teachers do not believe that their students can be reflective learners they are unlikely to act in ways that facilitate it or even to notice it when it occurs. The examples of reflective thinking presented in this chapter have thus been selected to include a young child and a mixed ability class.

4 Learning Rational Number Concepts

Rational number concepts include fractions, ratio and proportional reasoning and have been recognised as among the most difficult of all mathematical concepts for students to learn. Lamon (2007, p. 629) described these topics as arguably, "the most protracted in terms of development, the most difficult to teach, the most mathematically complex, the most cognitively challenging, the most essential to success in higher mathematics and science".

In relation to fraction representations of rational numbers, difficulties include understanding fractions as numbers and that the various equivalent representations of a given fraction all convey the same relationship between the single numbers involved in the notation of the number (Lamon, 2007). A particular instance of this is the alternative representations of fractions greater than one; as improper fractions (e.g., $^4/_3$) or mixed numbers (e.g., $1^1/_3$). Van de Walle, Karp, and Bay-Williams (2010) argued that fractions greater than one, including both representations of them, should be introduced throughout fraction learning rather than being treated as a separate topic.

Clarke and Roche (2009) cited research that highlighted the importance of students developing intuitions about the sizes of fractions and the difficulty that many have in so doing. There is evidence that students develop strategies for comparing the sizes of fractions without explicit teaching and that these strategies vary in their effectiveness in terms of their likelihood of resulting in correct comparisons (Clarke & Roche, 2009). Strategies with greater likelihood of success involve working flexibly with the relationship between the numerator and denominator and the impacts of each on the size of the number (Cramer, Post, & Del Mas, 2002). For example, $^3/_5$ can be identified as greater than $^3/_7$ because the smaller denominator in $^3/_5$ indicates that these pieces are larger and a number of larger pieces will be greater than the same number of smaller pieces. Similarly $^3/_5$ is smaller than $^4/_5$ because the same denominators indicate that all pieces are the same size and hence more of them (a greater numerator) will indicate a larger number. Cramer *et al.* (2002) also described transitive strategies that involve comparing each of the fractions with another value or benchmark, often one or one half. These strategies involve thinking about how many more pieces need to be added or removed to make the benchmark value, comparing the sizes of those pieces and hence the relative sizes of the two fractions.

A further crucial concept relates to the role of the whole. It is only meaningful to compare fractions, as was done in the preceding paragraph, if we assume that each fraction relates to the same whole. However, in many real contexts the wholes to which fractions relate are not the same. Watson, Beswick, and Brown (2006) described the responses of students in Years 5-8 to a problem that asked whether it was possible for Mary to have spent more than John if Mary had spent ¼ of her pocket money and John had spent ½ of his. Approximately two-thirds of the 650 students surveyed were unable to answer in a way that showed understanding of the role of the whole in this comparison.

Everyday usage of the word "fraction" to mean a small part of something is a further source of confusion (Lamon, 2007). Related to this, Watson, Beswick, and Brown (2012) observed that some Year 5 and 6 students used ¼ to refer to any small interval on a number line. Similarly, the familiarity from an early age of many children with the idea of halving affords this fraction a special status and proficiency with halving does not necessarily translate to understanding of other simple and even related fractions (Gould, Outhred, & Mitchelmore, 2006).

5 Examples of Reflective Learning

Transcripts of three teaching and learning episodes are presented in the sections that follow. In each case student behaviours that evidence reflective activity are highlighted and task features and teacher actions that seemed to support the student's reflection are identified. Finally the knowledge and beliefs that appear to underpin the teachers' actions are described.

5.1 *Understanding one third: Year 2*

This teaching episode is from an individual interview conducted by a final year pre-service teacher, Esther, as part of her honours research project. The Year 2 student, Lizzy, was 7 years old. Esther was interested in the capacity of young children to understand fraction concepts that were more sophisticated than those mandated by the curriculum. In her interviews she began with ideas about half that Year 2 students could be expected to understand, then progressed to other fractions, including thirds, that were beyond curriculum expectations.

Esther: I've got two chocolate bars. Do you like these ones?

Lizzy: Yes.

Esther: They're good aren't they. If we had three children and only two chocolate bars, how would you share the chocolate bars out? Three children and only two chocolate bars.

Lizzy: Mm ...

Esther: You can use the paper as well. Would like some paper?

Lizzy: Yes please. (She makes two vertical lines on the paper.) Two chocolate bars (looks at the actual chocolate bars and thinks for several seconds).

Lizzy: Wait, if you cut both in half that leaves you with four pieces so it's still not even. So if you cut that one in half and that one in half (gesturing with the pencil on the actual chocolate bars), those three halves and then you have the other one ...

Esther: You'd have a leftover.

Lizzy: So, and with your, the leftover piece you could cut it into four quarters.

Esther: And would you have another leftover piece?

Lizzy: Yes, three quarters, cut that one into three quarters. So you'd have (draws the picture in Figure 1).

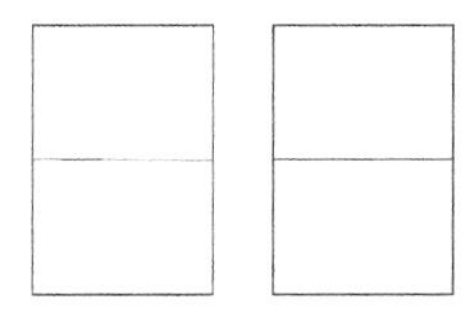

Figure 1. Lizzy's drawing showing 2 chocolate bars halved

You'd give one half to that person (makes a mark above the picture on the paper), then this half to another person (makes another mark). Then you'd do (adds lines so that the diagram looks like that shown in Figure 2),

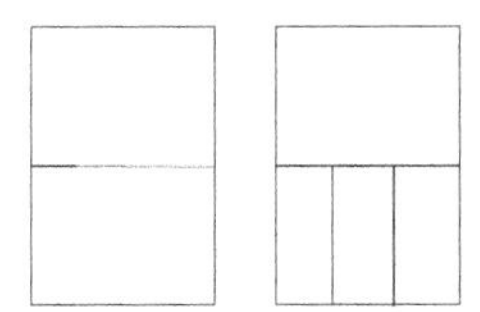

Figure 2. Lizzy's division of the leftover half

and give one quarter to each of them (makes marks next to each of three marks already on the page.

Esther: Is that a quarter?

Lizzy: Umm … (nods emphatically)

Esther: Well done, good job.

Later in the same interview the tasks focussed on understanding of one third. Lizzy spontaneously linked a paper folding activity to her division of the leftover half shown in Figure 2.

Esther: I'll fold this into thirds (folds an A4 page into thirds making creases to mark the thirds). Alright.

Lizzy: Wasn't that a third (pointing to the divided leftover half in her diagram – see Figure 2)

Esther: It was. Good. Well done. For thirds we need three equal sized pieces (unfolds the page).

Esther: Can you show me what one third would be? (Lizzy puts her finger on one of the sections). Two thirds? (Lizzy places her thumb on another section of the paper). And three thirds? (Lizzy spans all three sections with her hand). And what's three thirds?

Lizzy: (removes her had from the page) Umm ... a half. (looking at Esther) ... a whole.

Esther: Well done. It's a whole isn't it, a whole piece (gesturing to indicate the whole sheet of paper). I have ... (Taking ladybirds from a container in preparation for the next task).

Lizzy: It depends which way you do it. If you're doing two halves it's three quarters (pointing to her diagram shown in Figure 2).

Evidence of reflective thinking. Lizzy thought very hard about the task of sharing two chocolate bars fairly among three people. Evidence of this includes that she tried a strategy, halving, with which she was familiar and confident, but recognised that this would not on its own lead to a solution. Her diagram (Figure 2) shows that she realised that the fourth half in fact needed to be divided into three pieces but she appeared not to be aware of or comfortable with the word "third" instead referring to these pieces as quarters. This is consistent with the special status of one half (Gould *et al.*, 2006) and Watson, Beswick, and Brown's (2012, p. 162) report of the "common misuse of the word "quarter to mean a piece (often and small piece)". In the subsequent task that involved considering a page that had been folded into thirds, Lizzy was able to indicate one, two and three thirds but appeared still to have been thinking about the division of the left over half from the chocolate bar task because when she was asked what three thirds were her first response was one half. She was able to pick up cues from Esther's facial expression that prompted her to self-correct and say that the three thirds were one whole. Nevertheless, she was not satisfied to leave the matter there and simply to accept the fact that three thirds were one whole. Rather, she appears to have continued to work to reconcile the three parts of the half in the chocolate bar task with the three parts of the whole piece of paper. Her final statement in the transcript seems to be her attempt to verbalise the fact that she sees three pieces in the two contexts as different, as indeed they are. At the heart of her dilemma is that fact that in the chocolate bar task the 'whole' that was divided into thirds was in fact half a chocolate bar, whereas in the folded paper task, the 'whole' was the whole sheet of paper. The importance of the whole and the impact that varying it has upon the absolute size of a given fraction is a crucial concept that many students struggle with into the middle years of schooling (Watson, Beswick, & Brown, 2006).

Teacher actions that encouraged reflective thinking. A key contributor to Lizzy's reflection on these tasks was the fact that they were presented in the same teaching episode, although not consecutively. Although it is not clear that Esther was aware of the fact, encountering the idea of three equal parts in contexts in which the whole differed created a cognitive dilemma for Lizzy that she worked to resolve. Esther was deliberate in her choice of tasks that were likely to be challenging for a Year 2 student and that would provide opportunity to extend students' thinking to grapple with concepts that were new and at the edge of their ability to engage with meaningfully. Beyond the task choice the fact that Esther refrained from commenting on Lizzy's responses or correcting her use of terminology was likely to have helped Lizzy to feel able to articulate her thinking. Rather than evaluating Lizzy's solution to the chocolate bar task she used questions to prompt Lizzy to take on this role in relation to her own work. In addition, Esther's introduction of the term 'leftovers' also seemed instrumental in helping Lizzy to progress with the task.

Relevant teacher knowledge and beliefs. Esther's design of the interview and the tasks it contained arose from her belief that the curriculum was insufficiently challenging for many young students, and that young children can work with and develop understanding of fraction ideas beyond halving. The manner in which she conducted the interview suggests beliefs about teaching and learning mathematics that include that it is important not to evaluate students' responses but to encourage students to take on that responsibility for themselves; that questioning is useful to prompt reflection and to explore students' thinking; that having students struggle with challenging ideas is valuable; and that concrete materials can support students' thinking.

In terms of knowledge it is evident that Esther knew something about how children learn and how to engage them in tasks. She knew some things about how fraction understandings develop but there is also evidence that she may not have been aware of some key facts in this regard. For example, there is no evidence from the transcript that Esther understood that Lizzy was struggling with a problem that was essentially about the role of the whole and had not yet arrived at the understanding that one third referred to one of three equal parts of any whole. It is also possible, of course, that Esther was aware of this but chose not to deal with the issue at that time.

5.2 *Comparing fractions: Year 5*

The following example is from a one-on-one interaction between a teacher and a Year 5 student in which the student, Jotham, is trying to decide which of the juice mixes shown in Figure 3 would be strongest (most orangey).

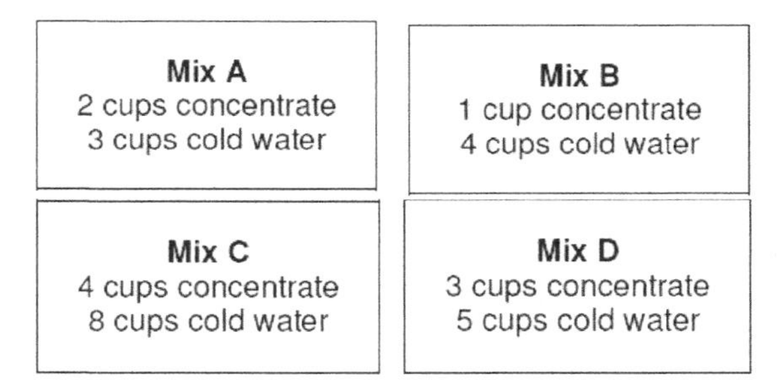

Figure 3. The four orange juice mixes

The point at which the transcript starts is 8 minutes into Jotham's efforts to solve the problem. He has established that the most orangey mix is either Mix A or Mix D and that he needs to determine which of $^2/_5$ and $^3/_8$ is the bigger fraction in order to decide between these. He had recognised that finding a common denominator would be helpful and, with some prompting, had been able to work out that 40 would be a common denominator. He had then tried intuitively to work out what the numerators would be and had said that $^2/_5$ would be about 15 of 16 fortieths because $^{20}/_{40}$ was half and $^{2}/_{5}$ was less than that. He could not see a way to say exactly how many fortieths $^{2}/_{5}$ was and so tried another approach detailed in the following transcript.

Jotham:	… I know.
Lyn:	What do you think? You're going to draw a diagram? (Jotham draws a circle and begins to mark radial lines on it). What's that a diagram of?
Jotham:	Pizza (completes the diagram to show 6 pieces).
Lyn:	How many pieces have you got there?
Jotham:	Five.
Lyn:	Have you got five?
Jotham:	(counts the pieces) No, I've got six. (draws another circle and divides it into five pieces of approximately equal size, scribbles over the first drawing, then draws another circle).
Lyn:	What are you drawing in this one?
Jothan:	Eight (draws radial lines to create eight approximately equal pieces). Three eighths, roughly that amount (shading 3 of the eight pieces).
Lyn:	Uh, Huh.
Jotham:	Two fifths, roughly that amount (shading two of the five pieces in the other circle).

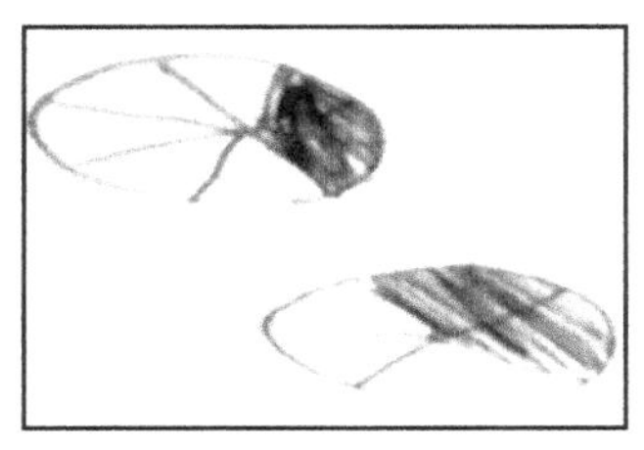

Figure 4. Diagrams for comparing $^{2}/_{5}$ and $^{3}/_{8}$

Jotham:	Yes, so that's (pointing at the circle divided into eight pieces) got still another eighth to go.
Lyn:	To make a half?
Jotham:	Yes … and this (pointing at the circle divided into five pieces) has got only about half a fifth.
Lyn:	So what fraction is half a fifth?

Jotham: Um … it's like this (makes the notation shown in Figure 5)

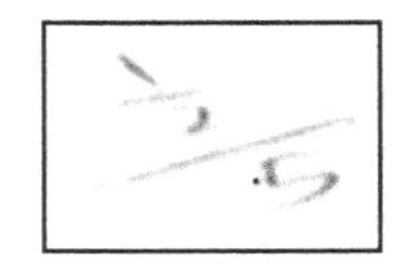

Figure 5. Notation for half of a fifth

Lyn: That's right, that is half a fifth. If you divided all the fifths in your picture in half, just imagine that happening, how many of them would there be? How many pieces?

Jotham: Ten.

Lyn: So what's half a fifth? What's one of your half fifths then?

Jotham: A tenth.

Lyn: It's a tenth, isn't it.

Jotham: Yes.

Lyn: So how much less than a half is your two fifths in your picture? How much more do you need to make a half?

Jotham: Tenth.

Lyn: A tenth, OK. And how much more do you need to make a …

Jotham: An eighth, that's bigger (pointing to the fraction, $^2/_5$)

Lyn: Which one's bigger?

Jotham: This one (points again to $^2/_5$).

Lyn: How do you know?

Jotham: Because, that's a tenth (pointing to the part of the diagram showing $^2/_5$ that would need to be shaded to make a total of a half shaded), OK, that's an eighth (pointing to an additional piece on the diagram showing $^3/_8$), and the smaller the numbers the bigger are the pieces.

Lyn: So the eighth is bigger than a tenth.

Jotham: Yes, and that one's (pointing to $^2/_5$) got less to go to make a half so Mix A.

Evidence of reflective thinking. Evidence that Jotham was reflecting on his work as he progressed with the task includes the fact that he modified his initial pizza diagram when prompted to check the number of pieces. Even before the episode presented in this transcript he had unsuccessfully attempted to use ideas related to common denominators to compare the fractions but had recognised that this was not working. Once he had produced the diagrams (see Figure 4) he acknowledged that they showed the two fractions only "roughly" and did not rely on their appearance to compare the fractions. Instead he used a transitive strategy described as effective by Cramer *et al.* (2002), comparing the area shaded in each case with one half and drawing upon existing knowledge about unit fractions: specifically that a larger denominator indicates a larger piece of the whole.

Teacher actions that encouraged reflective thinking. The task chosen by the teacher was challenging for Jotham but accessible. The teacher also refrained from intervening even though Jotham struggled with the problem for a total of more than 10 minutes. She could have helped him to check whether $^{2}/_{5}$ was in fact $^{15}/_{40}$ or $^{16}/_{40}$ but chose not to. Instead she allowed him to pursue another approach that was meaningful for him and that ultimately lead to the solution.

Relevant teacher knowledge and beliefs. Lyn seemed confident that Jotham could and eventually would solve the problem. Her decision not to help him work with common denominators is consistent with a belief in the value of allowing students to engage in mathematical reasoning that is meaningful to them. She appeared to value this above developing his understanding of common denominators despite the efficiency and broad applicability of this approach. Decisions about the length of time for which students should be allowed to struggle with a difficult problem draw upon knowledge of the relevant mathematics and how students typically learn it. This informs judgments about whether or not the student's efforts are likely ultimately to be useful in developing the desired mathematical understanding. They also rely on considerable depth of knowledge of the particular student, and his/her ability to persevere and to tolerate a degree of frustration. In this case, Lyn's decision not to intervene further in relation to common denominators led to Jotham using an approach that revealed his sound intuitions about fractions underpinned by an understanding of the proportional nature of equivalent fractions on which procedures involving common denominators rely. His pizza diagram approach also provided evidence that he understood the value of using benchmarks like one half to compare fractions, and that he seemed to appreciate the need for the pieces to be of equal area even though his drawings alone did not clearly show this. Lyn's decision may, therefore, have been driven at least in part by curiosity about Jotham's thinking and beliefs about the potential usefulness of insights into students' thinking.

5.3 *Understanding equivalent fractions: Year 7*

The following transcript is from a Year 7 lesson involving a mixed ability class and taught by a teacher, Andrew, with 25 years of experience. In this case the focus is on the activity of the class rather than on a single student. The lesson was intended to be about multiplication of fractions but Andrew quickly realised that he needed to adjust his plans and focus instead on representing mixed numbers as improper fractions. The lesson began with $2\ ^2/_3 \times 1\ ^1/_4$ written on the whiteboard.

Andrew: What's the answer?

Paul: $3\ ^2/_{12}$

Andrew: (to whole class) Check it on your calculator.

After a few seconds,

Andrew: No? I suppose I could let you find your own solution but we'd be here till Christmas.

Christine: Turn the wholes into fractions.
$2\ ^2/_3 = {}^8/_3$, and $1^1/_4 = {}^5/_4$

Andrew: How did you do it?

Christine struggles to articulate her thinking but manages to explain her answer well.

Stacey giggled at one point in Christine's explanation.

Andrew: One rule we have in this class; no one laughs at someone who's making an effort. (Andrew models Christine's solution with fraction tiles on the overhead projector.)

Andrew: What about $1\ ^2/_3$?

Paul: $1\ ^2/_3 = {}^6/_3$

Andrew: How did you get it?

Paul: One whole is three thirds, and two thirds is five thirds.
Oh! $^5/_3$

Andrew:	What about 1 ¾?
Susan:	$^{6}/_{4}$
Ellie:	$1\ ^{3}/_{4} = ^{7}/_{4}$. You just add the two numbers.
Ryan:	Does it work all the time?
Mary:	Wouldn't it only work if you had one?
Andrew:	Give me an example that doesn't work. (He pauses, and then writes $7\ ^{9}/_{10}$ on the board.)
Joseph:	$7\ ^{9}/_{10} = ^{79}/_{10}$
Andrew:	How did you do it?
Joseph:	(Comes to the front and writes on the board.) $7\ ^{9}/_{10} = ^{10}/_{10} + ^{10}/_{10} + ^{10}/_{10} + ^{10}/_{10} + ^{10}/_{10} + ^{10}/_{10} + ^{10}/_{10} + ^{9}/_{10}$
Drew:	You put 7 and 9 together to make 79.
Andrew:	Will it work all the time? What about 1 ¼?
Jason:	Wholes × bottom + top
Andrew:	Will it work all the time? What about $2\ ^{2}/_{3}$? (Writes it on the board.)
Jason:	$^{8}/_{3}$ (Andrew writes Jason's answer on the board.)
Andrew:	$4\ ^{3}/_{5}$? (Writes it on the board.)
Carly:	$4\ ^{3}/_{5} = ^{23}/_{5}$

Evidence of reflective thinking. In this example the actions of a range of students in the class provide evidence that the class was engaged in reflecting upon the problems posed and solutions suggested as the lesson unfolded. Evidence includes instances in which a student volunteers a solution without prompting. For example, when Andrew mentioned the possibility of being "here till Christmas" Christine suggested, "Turn the wholes into fractions". There were also examples of students self-correcting such as when Paul realised his error when prompted to explain how he found $1^{2}/_{3} = ^{6}/_{3}$.There were also attempts to generalise based on examples that had been presented. For example, having just seen that 1 ¾ = $^{7}/_{4}$, Ellie suggested that, "You just add the two numbers", and later in the lesson when fractions with whole number parts greater than one had been considered, Jason offered, "Wholes × bottom + top". In addition, there were several instances of students asking other students the kinds of questions that characterised Andrew's teaching, such as, "Does it work all the time".

Teacher actions that encouraged reflective thinking. In contrast to the previous examples the tasks in this case were routine and were not in fact the tasks that Andrew had intended the lesson to be about. Nevertheless, the examples of mixed numbers that he chose to be written as improper fractions were effective in eliciting false generalisations that could then be refuted. Specifically, Andrew began with an example in which the whole number part of the mixed number was one. These are the only cases in which adding the numerator and denominator of the fractional part does indeed result in the numerator of the appropriate improper fraction. His next example had ten as the denominator; the only situation in which juxtaposing the whole number and the numerator of the fractional part results in the numerator of the improper fraction. Finally he chose examples with whole number parts of two and four and the lesson culminated in the appropriate generalisation being reached.

In terms of managing the classroom discourse, it was apparent that the students were accustomed to engaging in conversation in which everyone's contribution was listened to, and in which explaining and justifying solutions, and attempting to generalise were expected (Beswick (2007) provides further examples and description of this class at work). An important contributor to the students' engagement in the class conversation was that Andrew responded in the same way regardless of whether a suggestion was correct, he permitted unsolicited contributions, and at times allowed the conversation to go on between the students without his involvement. His reprimand of Stacey who giggled at Christine's explanation exemplifies Andrew's role in establishing and maintaining a classroom environment in which thinking and respect were valued and hence students could engage publicly in reflective thinking.

It is instructive to consider Andrew's use of questions in light of the categories used by Hogan *et al.* (2012) in their analysis of classroom talk in Secondary Three mathematics classrooms in Singapore. The transcript of Andrew's lesson includes four performative questions, that is, questions focussed on eliciting a correct answer (e.g., What about $1^2/_3$?). There are also two procedural questions (e.g., How did you do it?) and two conceptual questions (e.g., Will it work all the time?). As Hogan *et al.* (2012) pointed out, conceptual questions are often associated with teaching for understanding but their analysis of Singaporean data demonstrated that other sorts of questions, such as performative and procedural, that are typically less well regarded in the literature promoting teaching for understanding can lead to conceptual talk and do not necessarily close down classroom dialogue. Andrew's classroom provides an Australian example of a possibly similar phenomenon. It is also worth noting that the physical arrangement of Andrew's classroom was not designed to facilitate group work. Rather pairs of desks were arranged in rows facing the front and the teacher and whiteboard were the focus of attention. Nevertheless, as was argued in Beswick (2007) the teaching emphasised mathematical thinking and the development of understanding.

Relevant teacher knowledge and beliefs. Andrew had been teaching for 25 years and had amassed considerable knowledge of how students learn in general and how they learn particular mathematical concepts. He had developed a repertoire of examples on which he could draw even when, as in the situation described here, he needed to change the focus of a lesson without warning. He was aware of students' capacity to overgeneralise and of misconceptions about fractions that they are likely to form and hence could anticipate these in his teaching.

As described by Beswick (2007), Andrew's practice was underpinned by beliefs that centred on the role of the teacher. Specifically, he believed that the teacher has a responsibility to take an active role in facilitating and guiding students' construction of mathematical knowledge, and to establish an orderly and mathematics focussed classroom environment. It was also evident that he believed that in such an environment students could learn from one another as well as from the teacher. Andrew's superficially traditional looking classroom was in fact centred on the development of mathematical thinking and understanding. It illustrates the point made by Askew *et al.* (1997), Watson and De Geest (2005), and Beswick (2007) that apparently similar surface level practices can be underpinned by quite different beliefs and it is the underpinning beliefs that in fact matter in terms of students' learning.

6 Reflective Learners and the Teacher Knowledge and Beliefs that Support Them

Collectively the three examples presented illustrate important characteristics of reflective learners. Reflective learners are focussed on and engaged with the task on which they are working; they think hard about the problem. In addition, they exercise metacognitive thinking in the process of working with the mathematics. Specifically, they: monitor, critique and modify as necessary the strategies that they try; self-correct when they become aware of an error in their reasoning or solution; take responsibility for judging whether the solution they obtain is satisfactory; seek feedback on their progress but do not expect to be excused from the work of thinking and learning; make connections with earlier tasks and draw upon existing understandings; and recognise the limitations of the strategies that they use. In supportive and safe classroom contexts they may volunteer solutions, articulate attempts to generalise, attend to one another's ideas as well as to the teacher, make arguments and defend viewpoints, acknowledge their own uncertainty, and ask questions. Often, struggling with unfamiliar concepts is accompanied by struggle with unfamiliar vocabulary and so students can find articulating their thinking difficult. Nevertheless, when they do there is much to be learned about their thinking by an attentive teacher whose aim is to explore his/her students' thinking.

The particular tasks on which the students described in this chapter worked were important in shaping the way in which they engaged with the relevant mathematics and the nature of the mathematical understandings that they might have developed as a result. Choosing appropriate, challenging yet accessible tasks that can facilitate intended mathematical thinking and learning (Muir, 2008) and then converting these into effective lessons are important parts of the teacher's role (Sullivan, Clarke, & Clarke, 2009). They draw upon teachers' knowledge of mathematics, the ways in which students typically learn specific aspects of mathematics including likely misconceptions and difficulties, and the mathematics curriculum.

In addition, the teachers in these examples had general pedagogical knowledge as well as pedagogical content knowledge particular to mathematics teaching that enabled the provision of appropriate materials to support students' thinking, and that informed their 'in the moment' decisions about appropriate wait time, and when to refrain from commenting, evaluating solutions, or endorsing particular approaches or procedures. They also knew how to model the use of appropriate vocabulary, to encourage students to take responsibility for judging the success of their own efforts, prompt students to explain their thinking, and ensure a sustained focus on important mathematical ideas. Inherent to their ability to do these things was knowledge of both the cognitive and affective characteristics of their students.

The ways in which teachers enact their knowledge, and indeed what they accept as known about mathematics and how to teach and learn it is shaped by the beliefs they hold about these things. At a fundamental, possibly unconscious, level what teachers believe mathematics actually is, what it means to know, learn and do mathematics, and why it is important underpins beliefs about appropriate ways to teach it. Teachers' beliefs about their own and their students' roles in the teaching and learning process are also fundamentally important and related to what they believe about their students. Beliefs about what students need to know, how they should learn, what motivates them, what they can do, and their capacities to learn mathematics are especially influential.

If teachers are to nurture reflective learners the following beliefs seem, from the examples discussed in this chapter, to be particularly relevant. Firstly, that mathematics is about sense making, and that depth of understanding is more important than getting through the curriculum content. In this context understanding is taken to mean, "insight, comfort, and flexibility in dealing with rational numbers" (Lamon, 2002, p. 636). Teachers who hold such beliefs will not be satisfied with students simply memorising and applying procedures but will also want them to think carefully about and understand the reasons for which procedures work, their mathematical bases, and how the particular mathematics involved relates to other aspects of the discipline. For students to achieve these things they must be reflective learners and teachers who believe in the importance of understanding will strive to develop reflective learners. Secondly, teachers who nurture reflective learners will necessarily be interested in their students' thinking and believe that inquiring into it is central to their ability to help them to learn mathematics meaningfully. As a result they will prompt students to explain their thinking and examine and evaluate alternative strategies. Again these things characterise reflective learners. Thirdly, teachers need to believe in the capacity of their students to learn mathematics. Finally teachers need to believe in the importance of their role and influence; that they are responsible for creating and maintaining a learning environment in which students have the time and encouragement to be thoughtful reflective learners and where there is an expectation that everyone will engage in thinking deeply about mathematics.

7 Conclusion

The examples in this chapter are intended to illustrate student behaviours that are indicative of reflective learning and to examine the teacher actions that appear to support these behaviours. At a more fundamental level it is contended that teachers' actions are shaped by their beliefs about the nature of mathematics and how it can best be taught and learned. In addition, teachers' knowledge of pedagogy, both general and specific to mathematics and of their students and the curriculum constrain the range of actions available to them.

The examples can assist teachers to think about what reflective learning of mathematics looks like and hence to be able to identify it in their students. Although teacher actions that appear to have supported the students' reflective activity are identified in each case it is important to stress that nurturing reflective learners cannot be reduced to a list of specific things to do or not do. Such a simplistic approach would be unlikely to be effective for at least two reasons. First, practices that are effective in developing reflective learning in one context or with a particular student may not be effective in another context or with other students; part of the knowledge that teachers need relates to the specific students and student groups with whom they work. Secondly, the effectiveness or otherwise of teaching practices appears to be dependent upon the teachers' purpose in using them (Askew *et al.*, 1997; Beswick, 2007; Watson & De Geest, 2005). Teacher purposes relate to the beliefs and it is these that represent a fundamental driver of teacher practice and hence student learning.

In attempting to identify the beliefs that underpinned the teaching described in the examples presented in this chapter it is acknowledged that the process is necessarily inferential and hence uncertain. In each of the cases additional information gleaned from talking to the teacher and/or observing his/her teaching on other occasions informed these inferences but they can still not be regarded as in anyway certain. It is, however, hoped that these examples will prompt teachers to examine their own practice and to reflect upon the reasons for which they do the things that they do. All teachers of mathematics have beliefs about what mathematics is and what it means to know, do and learn the subject. Even if they are subconscious these beliefs will have a profound impact on teaching (Beswick, 2012). Deep and lasting change in practice requires teachers to rethink their beliefs about such fundamental issues and to do the cognitive work of thinking through the implications of what is believed about mathematics for the ways in which they teach it, including how they perceive their own role, and envisage learning to occur. Similarly, beliefs about student capacity to learn, either as individuals or as classes, can shape teaching in powerful ways (Beswick, 2004). The examples presented here were deliberately chosen to include a young student and a mixed ability class to illustrate that reflective learning is possible for young students and for students of a range of abilities. Again being aware of one's beliefs and thinking carefully about the bases of them and impacts that they have is an essential step in changing practice.

The examples also illustrate the influence of teachers' knowledge on their ability to foster reflection in students. Teachers need to draw upon the full range of knowledge types listed by Shulman (1987). Without relevant knowledge the implementation of the best intentions are constrained. In summary, in order to nurture reflective mathematics learners, teachers themselves need to be reflective students of their own practice, and of the beliefs and knowledge that underpin that practice.

Acknowledgement

The author would like to thank the students and teachers who participated in the teaching episodes presented in this chapter.

References

Askew, M., Brown, M., Rhodes, V., Johnson, D., & Wiliam, D. (1997). *Effective teachers of numeracy: Report of a study carried out for the Teacher Training Agency*. London: King's College, University of London.

Ball, D.L., Thames, M.H., & Phelps, G. (2008). Content knowledge for teaching: What makes it so special? *Journal of Teacher Education, 59*(5), 389-407.

Beswick, K. (2004). The impact of teachers' perceptions of student characteristics on the enactment of their beliefs. In M.J. Hoines & A.B. Fuglestad (Eds.), *Proceedings of the 28th annual conference of the International Group for the Psychology of Mathematics Education* (Vol. 2, pp. 111-118). Bergen: Bergen University College.

Beswick, K. (2005). The beliefs/practice connection in broadly defined contexts. *Mathematics Education Research Journal, 17*(2), 39-68.

Beswick, K. (2007). Teachers' beliefs that matter in secondary mathematics classrooms. *Educational Studies in Mathematics, 65*(1), 95-120.

Beswick, K. (2011). Knowledge/beliefs and their relationship to emotion. In K. Kislenko (Ed.), *Current state of research on mathematical beliefs XVI: Proceedings of the MAVI-16 conference June 26-29, 2010* (pp. 43-59). Tallinn, Estonia: Institute of Mathematics and Natural Sciences, Tallinn University.

Beswick, K. (2012). Teachers' beliefs about school mathematics and mathematicians' mathematics and their relationship to practice. *Educational Studies in Mathematics, 79*(1), 127-147.

Chick, H., Pham, T., & Baker, M.K. (2006). Probing teachers' pedagogical content knowledge: Lessons from the case of the subtraction algorithm. In P. Grootenboer, R. Zevenbergen, & M. Chinnappan (Eds.), *Identities, cultures and learning spaces: Proceedings of the 29th annual conference of the Mathematics Education Research Group of Australasia* (pp. 139-146). Adelaide: MERGA.

Clarke, D.M., & Roche, A. (2009). Students' fraction comparison strategies as a window into robust understanding and possible pointers for instruction. *Educational Studies in Mathematics, 72*, 127-138.

Cramer, K.A., Post, T.R., & del Mas, R.C. (2002). Initial fraction learning by fourth- and fifth-grade students: A comparison of the effects of using commercial curricula with the effects if using the Rational Number Project Curriculum. *Journal for Research in Mathematics Education, 33*(2), 111-144.

Gould, P., Outhred, L., & Mitchelmore, M. (2006). One-third is three-quarters of one-half. In P. Grootenboer, R. Zevenbergen, & M. Chinnappan (Eds.), *Identities, cultures and learning spaces: Proceedings of the 29th annual conference of the Mathematics Education Research Group of Australasia* (Vol. 1, pp. 262-269). Adelaide: MERGA.

Green, T.F. (1971). *The activities of teaching*. New York: McGraw-Hill.

Hogan, D., Rahim, R.A., Chan, M., Kwek, D., & Towndrow, P. (2012). Understanding classroom talk in secondary three mathematics classes in Singapore . In B. Kaur & T.L. Toh (Eds.). *Reasoning, communication and connections in mathematics* (pp. 169-197). Singapore: World Scientific.

Lamon, S.J. (2007). Rational numbers and proportional reasoning: Toward a theoretical framework for research. In F.K. Lester Jr. (Ed.), *Second handbook of research on mathematics teaching and learning* (Vol. 1, pp. 629-667). Charlotte, NC: Information Age Publishing.

Muir, T. (2008). Principles of practice and teacher actions: Influences on effective teaching of numeracy. *Mathematics Education Research Journal, 20*(3), 78-101.

Shulman, L.S. (1987). Knowledge and teaching: Foundations of the new reform. *Harvard Educational Review, 57*(1), 1-22.

Sullivan, P., Clarke, D., & Clarke, B. (2009). Converting mathematics tasks to learning opportunities: An important aspect of knowledge for mathematics teaching. *Mathematics Education Research Journal, 21*(1), 85-105.

Van de Walle, J., Karp, K.S., & Bay-Williams, J. (2010). *Elementary and middle school mathematics: Teaching developmentally*. Boston: Pearson.

Watson, J.M., Beswick, K., & Brown, N. (2006). Teacher's knowledge of their students as learners and how to intervene. In P. Grootenboer, R. Zevenbergen, & M. Chinnappan (Eds.), *Identities, cultures and learning spaces: Proceedings of the 29th annual conference of the Mathematics Education Research Group of Australasia* (Vol. 2). Adelaide: MERGA.

Watson, J.M., Beswick, K., & Brown, N. (2012). *Educational research and professional learning in changing times: The MARBLE experience*. Rotterdam: Sense.

Watson, A., & De Geest, E. (2005). Principled teaching for deep progress: Improving mathematical learning beyond methods and materials. *Educational Studies in Mathematics, 58*(2), 209-234.

Chapter 5

Metacognitive Reflection at Secondary Level

WONG Khoon Yoong

Metacognition is one of the five components in the Singapore mathematics curriculum considered as important to facilitate students' ability to successfully solve mathematics problems. It is one form of reflection that students are asked to engage in, usually at the end of a lesson or a topic. In this chapter, I will discuss theoretical underpinnings of metacognitive reflection, cite some Singapore studies about two aspects of metacognition at the secondary level, and discuss techniques that teachers might experiment with in their mathematics lessons. Teachers ought to consider research evidence and professional advice and find out for themselves which techniques "work" for their students so that the latter can become more effective mathematics learners. This will lead to better learning outcomes in the cognitive, affective, and social domains.

1 Introduction: Two Aspects of Metacognition

Many attempts have been made to help students become better mathematics problem solvers and mathematics learners because mathematics is required for future study in many subjects and productive employment in the 21st century. The Singapore mathematics curriculum framework includes *metacognition* as one of the five components to help students become better problem solvers and learners. This framework is common to all the mathematics syllabuses from primary to pre-university levels. It defines metacognition in terms of "monitoring of one's own thinking" and "self-regulation of learning" (Ministry of Education, 2012). These two aspects, namely monitoring of one's own thinking especially during problem solving and self-regulation of learning, are related to several meanings of *metacognition* and *reflection* found in the literature on problem solving and learning. Hence, the term *metacognitive reflection,* though not widely used in the literature, is adopted here to explore its roles for mathematics instruction at secondary level.

Flavell (1976) coined the term *metacognition* and defined it as

> one's knowledge concerning one's own cognitive processes and products ... active monitoring and consequent regulation and orchestration of these processes in relation to the cognitive objects or data ... in the service of some concrete goal or objective. (p. 232)

Over the past few decades, this term has taken on wider meanings and has become an important factor in the learning of many subjects. In mathematics, Schoenfeld (1987) identified three aspects of metacognition to be self-awareness, control, and belief about one's cognition. Fogarty (1994) used the term *metacognitive reflection* to cover planning, monitoring, and evaluating of learning carried out by the learners. In more recent years, reflection is linked to assessment, giving rise to the notion of *reflective assessment* in mathematics (e.g., Ellis & Denton, 2010). In a nutshell, *metacognitive reflection* includes elements of metacognition (such as awareness and regulation) and reflection (as in looking back in the Polya's model of problem solving). In this chapter, the terms *metacognition* and *metacognitive reflection* will be used interchangeably.

The most recent Singapore mathematics curriculum document emphasises the importance of cultivating the habit of reflection from an early age to support "the development of metacognition" (Ministry of Education, 2012, p. 25). Overseas reviews of research about how people learn also support the importance of metacognition. For example, in a widely cited report, the National Research Council (2005) in the United States has included self-monitoring or metacognition as one of the three overarching principles of how students learn mathematics. The following sections build on these writings and highlight the role of metacognition in problem solving and learning. Although these ideas are applicable to all levels of schooling, only research about secondary students will be discussed here.

2 Metacognition During Problem Solving

This section examines metacognition during problem solving. The main goal of this aspect of metacognition is to help all students to improve their skills in solving a variety of mathematics problems. The next section compares metacognitive processes with cognitive processes during problem solving.

2.1 *Metacognitive processes and metacognitive questions*

During mathematics problem solving, two types of thinking are involved: cognitive processes and metacognitive processes. Metacognitive processes can support cognitive processes as indicated in Figure 1. Some non-metacognitive behaviours are given to contrast them with the productive ones.

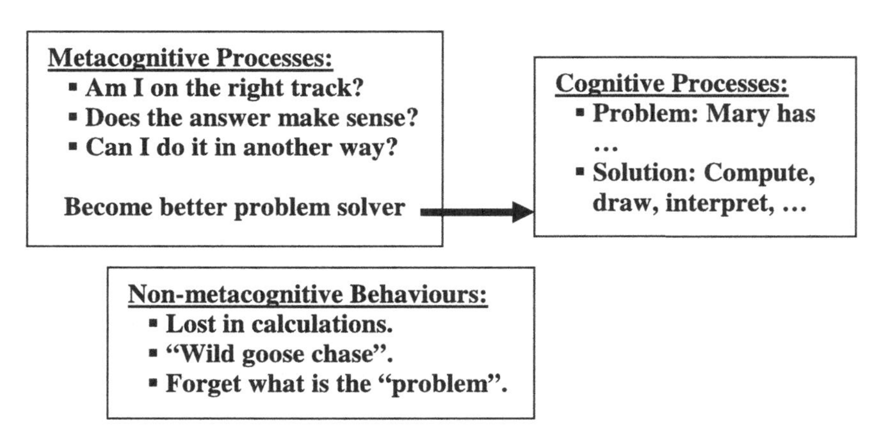

Figure 1. Metacognition during problem solving

The following problem is used to show the distinctions between these two processes.

> In 2010, the population of a small city was two million. Between 1910 and 2010, the population increased by 132%. Calculate the population in 1910.

Cognitive processes include specific mathematical procedures such as computations as well as generic processes such as relating to previous examples and justifying intermediate steps. The main cognitive processes for this solving this problem are deciding to calculate (2,000,000)/(2.32) and then completing the calculation. These cognitive processes are the main items stated in mathematics syllabuses.

On the other hand, metacognitive processes refer to monitoring intermediate steps by asking oneself questions such as "Am I on the right track?", "Does the intermediate answer make sense?" and "Does the approach I am taking look 'right'?" For this example, some specific metacognitive questions include:

- What does "increase *by*" mean? Is it the same as "increase *to*"? [Here, the student is monitoring his or her interpretation of the key word *by*.]
- Can this "increase by" be greater than 100%?
- Will a diagram be helpful?
- A value obtained by using a calculator is 0.862068; how many significant figures are meaningful for this problem? [Here, the student is reflecting on the reasonableness of an answer.]
- How can I check my answer?

It is believed that students who ask themselves such metacognitive questions will become better problem solvers because they will avoid pitfalls such as forgetting what is the "problem" to be solved, becoming lost in computations, or changing the cognitive processes mindlessly as described by Schoenfeld (1985) as "wild goose chase." Students should be trained in metacognitive monitoring so that they may be able to proceed when they are stuck. A popular approach is to embed metacognitive questions into a checklist, covering Polya's stages of problem solving (Polya, 1957). Figure 2 is an example of such a checklist. It includes emotions as part of metacognitive reflection. Teachers may use fewer questions to suit the experience of their students.

Problem Solving Checklist

Problem:

Insert problem here.

<table>
<tr><th>Steps and Prompts</th><th colspan="2">Write down your responses below.</th></tr>
<tr><td>0. Accept the problem.
a) Can I make sense of the storyline?
b) Is it an interesting problem?</td><td colspan="2">Describe your feelings about this problem.</td></tr>
<tr><td>1. Understand the problem.
a) What do I have to find?
b) What information is given?
c) Draw a diagram?
d) Write some equations?</td><td colspan="2">Re-write the problem in your own words.</td></tr>
<tr><td>2. Devise a plan.
a) Use method for a similar problem?
b) Try systematic guess and check?
c) Solve equations?
d) Other strategies/heuristics.</td><td colspan="2">Describe your plan in your own words.</td></tr>
<tr><td>3. Carry out the plan and monitor progress.
a) Am I systematic?
b) Re-read the problem to check whether I am answering the question.
c) Have I used all the relevant information?
d) If stuck for too long, change plan, ask for help, take a break, etc.</td><td colspan="2">Show your working. (Optional) Jot down comments and observations that come to mind.</td></tr>
<tr><td rowspan="2">4. Look back and learn.
a) Is the answer reasonable?
b) Does it satisfy the conditions and information?
c) Can I solve it differently?
d) How did I feel about solving this problem?
e) Can I make up a similar problem and solve it?
f) What do I need to remember about this type of problems?</td><td>Check answer.</td><td>Feelings.</td></tr>
<tr><td>Another method.</td><td>Make up similar problem and solve it.</td></tr>
</table>

Figure 2. Problem solving checklist based on Polya's stages

2.2 *Local studies about problem solving behaviours*

Four local studies (conducted in Singapore) are reviewed below. The first two studies had measured students' mathematics achievement after an intervention that included elements of metacognition, whereas the last two studies were self-reports of problem solving and metacognitive behaviours by students. Thus, local research to evaluate the impacts of metacognition on achievement is still very limited.

Metacognitive Instructional Strategies (MIS). For his doctoral study, Lee Ngan Hoe (2008) designed this MIS curriculum based on the use of a "problem wheel" or "reasoning wheel" proposed by Richard Paul. This wheel guides students through five inter-related steps using specific prompts: (a) what is given; (b) what to find; (c) draw a picture; (d) what topic is related; and (e) what formula can be used to solve the problem. This framework was used in conjunction with four other approaches: (a) writing a mathematics log; (b) effective questioning; (c) identifying structural properties of problems; and (d) pair or group problem-solving.

Lee conducted his study with four Secondary 1 Normal (Academic) classes using the quasi-control-experiment, pre-post-test design, with one experimental class and three comparison classes. The experiment was conducted over ten sessions. After solving a problem, the students answered five metacognitive questions, for example, "What are the questions that you asked yourself when you solved the problem?" This intervention resulted in several positive outcomes, including better general self-concept, mathematics self-efficacy, and problem solving scores. This study provides support for a metacognitive approach, but it was implemented in complex fashion together with other techniques. Hence, it is difficult to unravel the specific contribution of metacognition relative to other factors to explain the positive outcomes.

Mathematical Problem Solving for Everyone (M-ProSE). This was an intervention study to help students solve problems guided by Polya's heuristics and stages (Dindyal, Tay, Toh, Leong, & Quek, 2012). The researchers designed a set of problem solving "practical worksheets" to lead secondary students through Polya's stages in ten one-hour sessions (Toh, Quek, Leong, Dindyal, & Tay, 2011). For each problem, the student worksheet was scored in this way: the first three Polya's stages with evidence of cycling through the stages (10 marks); use of heuristics (4 marks); check and extend (6 marks). Scoring was considered an important part of this problem solving module because the researchers were "convinced that a root cause for the lack of success of learning problem solving is that problem solving is not assessed as part of summative assessment and national examination" (Dindyal et al., 2012, p. 5). One round of intervention was completed in 2009 with a Secondary 3 class. One of the researchers taught the first cycle using the practical worksheets and a school teacher taught the second cycle. Subsequently all the Secondary 2 classes in the school had to take this module.

At the end of the study, the students (n = 159) solved a given problem and completed a practical worksheet. About 70% of the students scored the maximum in the first three stages and 75% scored maximum in the use of heuristics. Stage 4 was more problematic: although 90% got at least 1 out of 3 marks, only 7.5% scored the maximum mark (6) for this stage. This shows that secondary students could learn to use this practical worksheet, but at the moment, there is no data about whether students who have undertaken this practical approach will become better problem solvers in various topics and more metacognitive in their problem solving approach. This requires further research.

Rahim, Hogan, and Chan (2012) reported some findings of the Core 2 project funded by the Centre for Research in Pedagogy and Practice at the National Institute of Education, Singapore. The representative sample included 1438 Secondary 3 mathematics students. Three 5-point Likert items were used to measure students' metacognitive knowledge, in terms of the extent to which they reported that their teacher "says we must try to understand how we learn ... to be aware of the strategies we use in solving problems ... to check if our answers make sense" (*ibid*, p. 19). Surprisingly, the students had the highest mean score about this kind of metacognitive knowledge (3.59) compared to their procedural knowledge (3.48), epistemic knowledge (3.36), and conceptual knowledge (3.29). Metacognitive knowledge was found to mediate between epistemic knowledge and procedural knowledge. The authors especially welcomed this "*relative* [original italics] strength of metacognitive knowledge" (p. 18). Thus, some mathematics teachers are already helping secondary students to use metacognitive strategies to complete instructional tasks.

Developing the Repertoire of Heuristics for Mathematical Problem Solving (MPS). Part of this study investigated the heuristics used by upper primary and lower secondary students to solve mathematics problems (Wong & Tiong, 2006). The heuristics used by about 140 Secondary 1 students were classified as: (a) systematic listing; (b) guess and check; (c) solve equation; (d) logical argument; and (e) diagrams that reflect the structure of the problem rather than simple representation of the problem context. The success of heuristic use by these students varied with the problems. However, some students did not use the same heuristic to solve parallel questions in March and in August, and this raises the issue about the pros and cons of consistency in use of heuristics in problem solving (Wong, 2008). The students also answered an open-ended questionnaire about their problem solving behaviours. The questionnaire consisted of nine items taken from the work of Callahan and Garafalo (1987). Some general findings from the questionnaire are:

- About 60% mentioned trying to understand the problems in general terms.
- Only about 15% wrote that they tried to check their work.
- About two thirds attributed their mistakes to carelessness, but only 8% attributed them to lack of understanding. Some of them mentioned the need to write more slowly and to avoid mental calculation.
- Describing reasons why they might not be successful in solving some of the problems, about 50% mentioned having not done the problems before; 25% not understanding the problems; 20% not good at the kind of problems given.
- Suggesting ways to improve their problem solving performance, about 60% mentioned practice in brief and generic terms.

Many of the written comments were quite brief. There could be three reasons for this. First, these Secondary 1 students had weak metacognition, i.e., they were not keenly aware of their own thinking when they engaged in problem solving. This is consistent with a claim made by Clarke (2006) that "as much as 90 percent of our learning and problem solving may be automated and unconscious" (p. 4). The second reason is that these students lacked the ability to write about their mental events because of weak English competence. Finally, they were not highly motivated to write extensively to answer the open-ended questions in the survey. If the first explanation about weak metacognition were true, then the challenge to raise the level of metacognitive awareness among students is a tough one. This has to be based on effective ways to teach metacognition.

2.3 *Teaching metacognition*

Two broad approaches for teaching metacognition are through the use of checklists and through modelling the process of solving unfamiliar problems upfront. These two approaches are addressed below, while other teaching ideas can be found in Wong (2002).

Use of checklists. As noted above, getting students to use a checklist of reflection questions is a very popular approach, such as the one in Figure 2. These questions can also be written in cards to enhance scaffolding (Holton & Clarke, 2006). First, teachers should ensure that they can answer these reflection questions for the problems to be solved. It might happen that the reflection questions are so general that the students do not know what to write down as their "answers". This reduces the effectiveness of the checklist.

Next, the teacher has to model how to use the checklist. The "I do – We do – You do" instructional modelling procedure is effective. It includes the following steps, using a succession of problems:

- **I do**. Teacher demonstrates how to use the checklist with problem 1 by thinking aloud; this is different from just explaining the items in the checklist.
- **We do**. Teacher and pupils together work through problem 2; teacher guides students through the checklist with probing questions, engaging the students in discussion, and referring back to key features in problem 1, if necessary.
- **You do**. Students work independently or in pairs on problem 3. Teacher walks around to make sure that the students are on task and to jot down any weaknesses in the use of checklist that need to be discussed with the whole class later on.

This process is consistent with Vygotsky's approach of providing scaffolding by more knowledgeable others, followed by fading (Wood & Wood, 1996), so that the students gain mastery eventually. This 3-step process needs to be repeated until the students become proficient in using the checklist. At this stage, teachers are encouraged to reflect on their experience of teaching in this way:

- Which items in the checklist are very helpful to the students? Which items are not?
- How well can you help students apply the helpful items?
- If you collect these reflection checklists from the students, how would you use them as assessment *for* learning?

Solve unfamiliar problems upfront. Teachers often solve the problems before they assign them to the class. This hides from the students the valuable but uncomfortable experience of how one can recover from a "stuck" situation during problem solving. The teacher can demonstrate this recovery process by verbalising his or her thinking aloud when he or she solves an unfamiliar problem in front of the class. Making this thinking visible can help students deepen their metacognitive reflection; indeed, the teacher may also learn something valuable about his or her own problem solving behaviours and pedagogic skills under this challenge. For this to happen, the learning environment must be based on trust and risk taking rather than on authority and the pressure to get the right answers only. Figure 3 describes areas where students may be stuck and suggests ways to help them become "unstuck". Mason, Burton, and Stacey (2010) advised learners to acknowledge that being stuck "is a healthy state because you can learn from it" (p. 56), and they described several ways to tackle this problem at the entry, attack, and review phase of problem solving. Teachers need to modify the suggestions given in Figure 3 according to the specific problems and students who have different experiences with metacognitive prompting.

Polya's Stages	"Stuck"	"Unstuck"
0. Motivation.	▪ Not interested to tackle problem.	▪ Re-frame problem, different representations.
	▪ Too difficult; I am dumb.	▪ Use your brain; it grows (Dweck, 2006).
1. Understand the problem.	▪ Difficult key words/concepts; too many words I do not understand.	▪ Explain again in different ways or representations. ▪ Ask students to look up notes, textbooks, dictionaries etc.
	▪ Don't know what to find.	▪ Read aloud. Say it in your own words.
2. Devise a plan.	▪ Problem does not look "familiar". Don't know how to start ("stuck").	▪ Provide hints; give a simpler but more similar problem. ▪ Re-state the problem in goal-free format. What can you deduce? ▪ Label the problem: I am stuck because
	▪ Look "familiar" but cannot recall the formula/skill correctly.	▪ Ask probing questions that lead to a formula. What does this remind you of? ▪ Discuss with your neighbour. ▪ Solve a different problem, let mind digress, then come back to problem. ▪ Misled by "similar" problems!
3. Carry out the plan and monitor progress.	▪ Cannot complete the operations correctly.	▪ Practise with simpler problems.
	▪ Not systematic, incomplete steps here and there.	▪ Make a table, jot down the points; be neat and systematic.
	▪ Jump around aimlessly.	▪ Use checklist.
	▪ Give up too quickly; lack confidence.	▪ Think of similar past experiences, how long do you persevere before giving up?
4. Look back and learn.	▪ Do not check work and answer.	▪ Model process and make it a requirement.
	▪ Alternative methods.	▪ Solve-Pair-Share. Which method is "optimal" for you? ▪ This is *the* way vs. this *could* be a way

	▪ Extend the problem.	▪ Train students to do so; change numbers, context, wording, structure.
	▪ Feelings.	▪ Encourage and "celebrate" success. ▪ Label the emotions.
	▪ Main learnings.	▪ Discuss special features.

Figure 3. Problem solving: Stuck and unstuck

The above discussion shows that metacognition can be at a level higher than content-related cognitive processes. Thus, students may have greater difficulty mastering and applying metacognitive processes than learning the contents. Similarly, it may demand more sophisticated pedagogical skills when one teaches metacognitive reflection compared to standard mathematical procedures. More research is needed to investigate these issues about the teaching and learning of metacognition during problem solving.

3 Equip Students to Regulate their Learning

This second aspect of metacognitive reflection refers to how students regulate their mathematics learning inside the classroom and outside it. During lessons, students must be aware that they should pay close attention to teacher's explanations, listen to questions and answers that go on in the lessons, carry out seatwork and class activities as instructed, and so on. When the lessons are over, they should monitor completion of homework, plan for revision, write their own notes, seek help from others when in doubt, set learning goals, and other similar learning activities. An example from Flavell (1976, p. 232) can be used to illustrate three aspects of metacognition during learning:

- Awareness: "I notice that I am having more trouble learning *A* than *B*."
- Control: "I should double-check *C* before accepting it as a fact."
- Action: "I had better make a note of *D* because I may forget it … ask someone about *E* to see if I have it right".

This "learn about their own learning" is based on research in "learning to learn" and study strategies. The National Research Council (2005) referred to this aspect of metacognition as helping "students learn to take control of their own learning by defining learning goals and monitoring their progress in achieving them" (p. 2).

In a new taxonomy based on Bloom's work, Marzano and Kendall (2008) placed this aspect of metacognition at level 5 just below the topmost level 6. They included four processes: specifying goals, process monitoring, monitoring clarity, and monitoring accuracy. These processes are about improving effectiveness of learning in general. What they mean for the special discipline of mathematics learning needs to be worked out in detail.

The "learn to learn" approach has not gained sustained interest among mathematics educators compared to language educators. Writing about helping students to become mindful in their learning, Buoncristiani and Buoncristiani (2012) believed that "students make poor choices about how they study because they are unaware of alternatives and their consequences" (p. 128). Thus, the first step is to understand what students report about how they study, followed by guiding them to use more effective learning strategies.

3.1 *Local studies about learning strategies in mathematics*

Doing homework provides the essential practice for students to master new skills, and homework is a hot research issue internationally. Results from TIMSS 2007 (Mullis *et al.*, 2008) show that about 42% of Singapore 8th graders completed mathematics homework at least 3 to 4 times per week, with more than 30 minutes used for each homework assignment. This is 15 percentage points more than the international average for students who were found to have "high" homework index. However, it is not clear from the TIMSS findings how students handle homework as part of their learning.

About two decades ago, the author conducted a large-scale survey (n = 1318) of the study behaviours of Secondary 1 to Junior College students (Wong, 1990). Of the 36 items in the survey, memorising formula had the highest mean score (4.2 on a 5-point Likert scale), and nearly 80% of these students reported using this as a very frequent study activity. However, 68% of all the students also tried to understand what they memorised, suggesting that memorising is not a simplistic technique as is often portrayed by some educators. Two decades later, the TIMSS 2007 data show that 72% of Singapore 8th graders reported using memorisation in at least half of the lessons (international average was 63%), compared to only 46% relating what is being learned to everyday life and 60% explaining their answers or working on problems on their own (Mullis *et al.*, 2008). While this shows that memorisation is still a popular activity among Singapore students, it is important to note that "memorisation can be achieved by various means: reciting a formula orally, writing it down several times, making up mnemonics, using practice, and employing mental imagery" (Wong, 1990, p. 569). This suggests that memorisation may be a convenient umbrella term covering both active and passive activities. In addition to memorisation, students in the 1990 study also reported that they often completed homework on time (mean, 4.1) and paid attention in class (mean, 4.1), but they seldom tried different methods of solving the same problem (mean, 2.3) and borrowed mathematics books from the library (lowest mean, 1.5).

Part of a pilot study, *Enhancing Mathematics Performance of Mathematically Weak Pupils (EMP)* (Wong & Quek, 2009) investigated the study environment and study behaviours of 80 Secondary 1 Normal students. About 60% of these students had own study room at home, and 40% took private tuition in mathematics. These students rated highly traditional study behaviours in terms of frequency and usefulness: pay attention to teacher when he or she explains things; listen to others explaining things in class; hand in homework on time; and ask for help when they do not understand. On the other hand, metacognitive behaviours, such as writing reflections about lessons in a journal, making their own notes, and planning their own schedule of studying mathematics were reported to be used only occasionally or to be of limited use. Thus, consistently over the past two decades, a majority of Singapore secondary students have used traditional and generic strategies rather than more "constructivist" ones when they study mathematics. A plausible reason is that they have not been exposed to alternative learning strategies specifically designed for mathematics.

Luo, Paris, Hogan, and Luo (2011) examined the roles of different learning goals with respect to other psychological factors in the learning of mathematics among 1697 Secondary 3 students from 39 schools. Of interest here are the four metacognitive self-regulation items, e.g., "I ask myself questions to make sure I understand what I study in my math class" (p. 168). On a 5-point scale, the mean score for this metacognitive scale was 3.47, a moderate value. It had low correlations with an online mathematics test (0.11) and the Primary School Leaving Examination (PSLE) score of the students (0.06), but moderate correlations of about 0.50 with mastery goals, self-efficacy, and engagement. Thus, metacognition as awareness of learning has strong links with other affective constructs but weak link to cognition. Thus, the potential benefits of metacognition to enhance achievement may have to be found through its effects on non-cognitive factors.

3.2 *Teaching self-regulation of learning*

Wiliam (2011) claimed that "[s]tudents can develop sufficient insights into their own learning to improve it" (p. 146). Despite the sketchy local research to support this claim, it is still prudent for teachers to guide their students to acquire helpful learning strategies to become more effective and independent learners.

One basic strategy is for students to keep a daily homework log. This log should include the homework topics, submission date, teacher comments, and follow up actions by the students. Ellis and Denton (2010), drawing on the benefits of record keeping in medical practices and their own work, suggested that getting students to fill in a record sheet of "the amount of time they spend on their homework, the number of pages they read etc." could help them improve in learning because they become "more aware of the need to study" (p. 120) and what to pay attention to. The more conscientious students may also keep a record of the mistakes they have made in various topics and the corrections, and then use this record to complete assignments and prepare for tests.

In the latest version of the mathematics curriculum, the Ministry of Education (2012) recommends that students at all levels reflect on their learning using concept maps and journals, even sharing these reflections through blogs. Concept mapping can be a useful tool for metacognitive reflection, but the author is not aware of any research about how Singapore secondary students learn mathematics using concept maps. Indeed, training students to use concept maps effectively requires careful planning; see the procedures documented by Jin and Wong (2011).

Student Question Card (Wong & Quek, 2009). Some students are afraid or do not know how to ask questions when they cannot follow certain parts of the lessons. To help them develop the skill and disposition to ask mathematics questions, the research team created a set of laminated cards that cover four types of mathematics-related questions: *Meaning* (e.g., What do you mean by ...?); *Method* (e.g., Can you show us how to do this problem in another way?); *Reasoning* (e.g., Why do you do that ...?); *Application* (e.g., How do we use this in everyday life?). After about 10 minutes of explanation or activity, the teacher pauses and invites students to ask questions, if any, by selecting from the given list of questions. This scaffolding was intended to be the initial step to help students eventually learn to ask their own questions. The mean scores (on a 5-point Likert scale) of students' responses from the two secondary classes taught by different teachers are given in Table 1.

Table 1

Students' Feedback on Student Question Card (means)

	Items	Class A	Class B
Q18:	The question cards were easy to use.	4.03	3.74
Q20:	I could understand the questions on the cards.	4.32	4.00
Q22:	This "question cards" method helped me to understand mathematics better.	3.45	3.16
Q21:	I would like my teacher to use this "question cards" method to teach mathematics.	3.18	3.11
Q19:	I enjoyed using these question cards to learn.	3.61	3.03

The two teachers used this technique differently in different lessons. They may call on students who flashed cards of a particular type of questions or who normally did not ask questions during lessons. Results in Table 1 show that these students held different opinions about the use of this pedagogic technique, suggesting that it is more suitable for some students than others. Although the technique is easy to use, this does not seem to lead to stronger desire to use it or enjoyment toward its use. It is worthwhile exploring this observation with further research.

4 Concluding Remarks

This chapter has examined two aspects of metacognitive reflection, namely monitoring of thinking during problem solving and learning behaviours. There is great interest among mathematics educators and teachers in the first aspect about awareness, planning, control, and monitoring of metacognitive processes during problem solving because these processes are thought to help students become better problem solvers. However, when the problems are routine and easy, there is no need to engage in these sophisticated metacognitive processes, or in the words of a student, having to answer metacognitive questions is "a hindrance for the easy ones" (Dindyal *et al.*, 2012, p. 13). The few local studies reviewed above show that Singapore secondary students are aware of these problem solving processes in a broad and general way, but there is no evidence that this general level of awareness is helpful, especially when they have to solve non-routine or challenging problems. New research has to be conducted to find such evidence. Toward this end, teachers may conduct action research or join lesson study groups focussing on this aspect of metacognition. These professional development activities provide the necessary opportunities for teachers to try something new in a small scale and to learn from the experimentation before they implement it in day-to-day lessons. This follows from the following insightful observation made by Bill Gates (1999):

> Most teachers have a great love of learning, and they'll get excited about anything that will help kids learn. What teachers don't want is to be thrown into something they have not had the opportunity to learn about and become comfortable with. (p. 388)

In contrast to problem solving, metacognitive reflection as self-regulated learning, as reviewed in section 3, has not attracted much attention among mathematics educators here and in other countries in recent years, although there are previous notable exceptions, such as the PEEL project (Baird & Northfield, 1992) and a project at the Shell Centre (Bell, Crust, Shannon, & Swan, 1993). Singapore secondary students tend to use traditional learning strategies such as memorise formulae, pay attention, and complete assigned homework. Although these strategies are desirable in general, some students may not have used them consistently. For example, teachers have complained about the difficulty of getting weak students to submit homework; many schools have to allocate a few marks as part of continuous assessment to coerce them to do so. Furthermore, students need to develop alterative learning strategies that can strengthen their ability "to learn to learn", as part of 21st century competencies. The crux of this aspect of metacognitive reflection is to educate students to take greater responsibility for their own learning. Finding effective pedagogies that can help the intended students to inculcate the skills of self-regulated learning and the habit to use it is an urgent challenge for both teachers and researchers.

Ultimately, efforts that combine theories, practices, and research should help to transform metacognitive reflection from an unrealised potential into a tangible, mainstream technique that will prepare students to become better mathematics learners and problem solvers.

References

Baird, J.R., & Northfield, J.R. (1992). *Learning from the PEEL experience*. Melbourne: Monash University.

Bell, A., Crust, R., Shannon, A., & Swan, M. (1993). *Awareness of learning, reflection and transfer in school mathematics: Summary report*. ESRC Project: R000-23-2329. Nottingham: Shell Centre for Mathematical Education, University of Nottingham.

Buoncristiani, A.M., & Buoncristiani, P. (2012). *Developing mindful students, skillful thinkers, thoughtful schools*. Thousand Oaks, CA: Corwin.

Callahan, L.G., & Garofalo, J. (1987). Metacognition and school mathematics. *Arithmetic Teacher, 34*(9), 22-23.

Clark, R.E. (2006). Not knowing what we don't know: Reframing the importance of automated knowledge for educational research. In G. Clarebout & J. Elen (Eds.), *Avoiding simplicity, confronting complexity: Advances in studying and designing (computer-based) powerful learning environments* (pp. 3-14). Rotterdam: Sense Publishers.

Dindyal, J., Tay, E.G., Toh, T.L., Leong, Y.L., & Quek, K.S. (2012). Mathematical problem solving for everyone: A new beginning. *The Mathematics Educator, 13*(2), 1-20.

Dweck, C.S. (2006). *Mindset: The new psychology of success*. New York: Random House.

Ellis, A.K., & Denton, D.W. (2010). *Teaching, learning, and assessment together: Reflective assessments for middle and high school mathematics and science*. Larchmont, NY: Eye on Education.

Flavell, J.H. (1976). Metacognitive aspects of problem solving. In L.B. Resnick (Ed.), *The nature of intelligence* (pp. 231-235). Hillsdale, NJ: Lawrence Erlbaum.

Fogarty, R. (1994). *How to teach for metacognitive reflection*. Palantine, IL: IRI/Skylight.

Gates, B. (with Hemingway, C.). (1999). *Business @ the speed of thought: Using a digital nervous system*. Ringwood, Victoria: Viking.

Holton, D., & Clarke, D. (2006). Scaffolding and metacognition. *International Journal of Mathematical Education in Science & Technology, 37*(2), 127-143.

Jin, H.Y., & Wong, K.Y. (2011). Assessing conceptual understanding in mathematics with concept mapping. In B. Kaur & K.Y. Wong (Eds.), *Assessment in the mathematics classrooms: Yearbook 2011, Association of Mathematics Educators* (pp. 67-90). Singapore: World Scientific.

Lee, N.H. (2008). *Enhancing mathematical learning and achievement of secondary one Normal (Academic) students using metacognitive strategies*. Unpublished doctoral dissertation, Nanyang Technological University, Singapore.

Luo, W., Paris, S.G., Hogan, D., & Luo, Z. (2011). Do performance goals promote learning? A pattern analysis of Singapore students' achievement goals. *Contemporary Educational Psychology*, *36*, 165-176.

Marzano, R.J., & Kendall, J.S. (2008). *Designing & assessing educational objectives: Applying the new taxonomy*. Thousand Oaks, CA: Corwin Press.

Mason, J., Burton, L., & Stacey, K. (2010). *Thinking mathematically* (2nd ed.). Harlow: Pearson.

Ministry of Education, Singapore. (2012). *O-Level Mathematics: Teaching and learning syllabus*. Singapore: Author.

Mullis, I.V.S., Martin, M.O., & Foy, P. (with Olson, J.F., Preuschoff, C., Erberber, E., Arora, A., & Galia, J. (2008). *TIMSS 2007 international mathematics report: Findings from IEA's Trends in International Mathematics and Science Study at the fourth and eighth grades*. Chestnut Hill, MA: TIMSS & PIRLS International Study Center, Lynch School of Education, Boston College.

National Research Council. (2005). *How students learn: Mathematics in the classroom*. Committee on How People Learn. Washington, DC: National Academy Press.

Polya, G. (1957). *How to solve it* (2nd ed.). New York: Doubleday & Co.

Rahim, R.A., Hogan, D., & Chan, M. (2012). The epistemic framing of mathematical tasks in Secondary Three mathematics lessons in Singapore. In B. Kaur & T.L. Toh (Eds.), *Reasoning, communication and connections in mathematics: Yearbook 2012 of the Association of Mathematics Educators* (pp. 11-55). Singapore: World Scientific.

Schoenfeld, A.H. (1985). *Mathematical problem solving*. Orlando, FL: Academic Press.

Schoenfeld, A.H. (1987). What's all the fuss about metacognition? In A.H. Schoenfeld (Ed.), *Cognitive science and mathematics education* (pp. 189-215). Hillsdale, NJ: Lawrence Erlbaum Associates.

Toh, T.L., Quek, K.S., Leong, Y.H., Dindyal, J., & Tay, E.G. (2011). *Making mathematics practical: An approach to problem solving*. Singapore: World Scientific.

Wiliam, D. (2011). *Embedded formative assessment*. Bloomington, IN: Solution Tree Press.

Wong, K.Y. (1990). How secondary students learn mathematics. In K. Milton & H. McCann (Eds.), *Mathematical turning points: Strategies for the 1990s* (pp. 566-572). Hobart: Australian Association of Mathematics Teachers.

Wong, K.Y. (2002, April). Helping your students to become metacognitive in mathematics: A decade later. *Mathematics Newsletter*, *12*(5). Available from http://math.nie.edu.sg/kywong/

Wong, K.Y. (2008). Success and consistency in the use of heuristics to solve mathematics problems. In M. Goos, R. Brown, & K. Makar (Eds.), *Navigating currents and charting directions: Proceedings of the 31st Annual Conference of the Mathematics Education Research Group of Australasia, Vol. 2* (pp. 589-595). Adelaide: MERGA.

Wong, K.Y., & Quek, K.S. (2009). *Enhancing Mathematics Performance (EMP) of mathematically weak pupils: An exploratory study*. (Unpublished Technical Report). Singapore: National Institute of Education, Centre for Research in Pedagogy and Practice. Available from http://repository.nie.edu.sg/jspui/handle/10497/2900/

Wong, K.Y., & Tiong, J. (2006). *Developing the Repertoire of Heuristics for Mathematical Problem Solving*: Student problem solving exercises and attitude (Unpublished Technical Report). Singapore: National Institute of Education, Centre for Research in Pedagogy and Practice. Available from http://math.nie.edu.sg/kywong/

Wood, D., & Wood, H. (1996) Vygotsky, tutoring and learning. *Oxford Review of Education*, *22*(1), 5-16.

Chapter 6

Reflecting on an *Excellent Teacher's* Video Recorded Mathematics Lesson: What Can We Learn?

LIM Chap Sam CHEW Cheng Meng

This chapter discusses how the authors used a video recorded lesson of a primary mathematics Excellent Teacher to promote reflective thinking among in-service teachers so as to improve their teaching practices. The data were collected from 18 Malaysian Chinese primary school teachers and 39 Singaporean primary school teachers. During the workshops, the participating teachers were shown a 10-minute video recorded lesson and asked to write their reflections about the lesson based on two main questions. For the first question, they showed similar reflections like the lesson was lively and pupils were happily engaged in the classroom activities. However, five major themes emerged from their reflections on the second question, namely group size, use of mini-whiteboard, structured worksheet, use of measuring instruments, and giving alternative teaching ideas. Further, the video recorded lesson enhanced the teachers' understanding of the mathematical ideas and helped them to generate new teaching ideas.

1 Introduction

The focus of this chapter is to discuss and share our experience of using a video recorded mathematics lesson to promote reflective thinking among in-service teachers with the aim of improving their teaching practices. We begin the chapter by defining the meaning of reflection and reflective thinking. Based on the review of related literature, we highlight the importance of promoting reflective thinking among pre-service and in-service teachers.

While the review of literature shows that research on promoting reflective thinking among pre-service teachers is well documented, studies involving in-service teachers are still relatively scarce. Hence this chapter focuses on promoting reflective thinking among in-service teachers. In view of the advantages of using digital video as a mode of observing critical events in a classroom lesson, we have employed this tool in this study. In this chapter we will explain how this process was carried out among two groups of teachers during two different workshops. We then analyse and compare their responses to identify common themes and differences.

2 Review of Literature

2.1 *Reflection and reflective thinking*

The concept of reflection in teaching stems from the work of John Dewey (1933) who defined reflection as the ''active, persistent and careful consideration of any belief or supposed form of knowledge in the light of the grounds that support it and the further conclusions to which it tends'' (p. 9). Dewey suggested that reflective thinking consisted of five phases: namely suggestions, intellectualisation, hypothesising, mental elaboration, and testing hypothesis. For Dewey, reflective thinking is an important pre-requisite for teachers to improve their teaching. To achieve that, the attitudes of open-mindedness, a sense of responsibility, and wholeheartedness or dedication are essential to the potential development of a reflective teacher.

Schön (1983) built upon Dewey's (1993) work and proposed that there are two types of reflection, namely reflection-in-action and reflection-on-action. Reflection-in-action refers to considering a practice as it unfolds while reflection-on-action focuses on looking back on a practice. The former emphasises the importance of teachers being aware of their decisions as they teach and it depends very much on a kind of tacit knowledge called knowing-in-action. An example of reflection-in-action will be: when a mathematics teacher is teaching a mathematical concept, say 'addition of two fractions', she finds that her pupils cannot add two fractions with different denominators correctly because they have not mastered the concept of equivalent fractions. Thus, upon reflection, the teacher alters her teaching plan to revise equivalent fractions with her pupils before proceeding with her lesson. The latter, reflection-on-action emphasises the importance of teachers reflecting back on and critiquing their practice. Thus, both types of reflections emphasise the importance of experiential learning. As, Tremmel (1993) argues the process that underlies knowing-in-action invokes "abilities as 'feeling', 'seeing', or 'noticing' what it is you are doing, then learning from what you feel, see or notice; and finally, intelligently, even intuitively, adjusting your practice" (p. 436).

Killion and Todnem (1991) extended the past and present timeframes of Schön's reflection to the future timeframe which they called reflection-for-action, that is the desired outcome to guide future action (cited in Harford, MacRuairc, & McCartan, 2010). That is, the reflective thinking process simultaneously includes past (reflection-on-action), present (reflection-in-action), and future (reflection-for-action) timeframes.

2.2 *Importance of reflective thinking for teachers*

Knowledge and experiences that teachers have gained during their pre-service teacher education programme while sufficient to meet the needs and demands of their present students, they still need to improve in their profession continuously through lifelong learning. Making reflections or reflective thinking is one kind of lifelong learning.

Reflective thinking that involves both reflecting in oneself and reflecting on others could be a great way to learn for teachers. Valli (1997) described reflective teachers as those who "can look back on events, make judgments about them, and alter their teaching behaviours in light of craft, research, and ethical knowledge" (p. 70). To enhance the learning process Zeichner and Liston (1996) (cited in Jay & Johnson, 2002) add that, reflective teachers must move beyond simple questions such as whether or not their practice is working, to asking how it is working and for whom. They also advocate for teachers to critically examine the inherent values in their practice as well as how their practice will lead to change, a commitment to quality, and respect for differences.

As well as looking at one's own practice, peer observation — that is observing and reflecting on others' teaching — is another way that enables teachers to see the problems from a different perspective. Richards (1991) conducted a peer observation project in which teachers would work in pairs and take turns observing each other's classes. The teachers reported that they gained new insights into aspects of their own teaching from their colleague's observations such as providing more detailed information on student performance during specific aspects of the lesson, revealing unexpected information about interaction between students, and providing useful information on the group dynamics that occur during group work. As a result of the information gathered from their colleague's observations, some teachers identified aspects of their teaching that they would like to change such as limited range of teaching strategies that they had been using, need to give students more time to complete some of the activities they used in a lesson, and need to develop better time management strategies. The peer observation project also benefited the teachers' department they worked by developing a better working relationship among colleagues and bringing out some broader issues about teaching and the programme during post-observation discussions. Additionally, the participating teachers reported that they would like to use peer observation on a regular basis.

Hence, as highlighted by Zeichner and Liu (2010) in their critical analysis of reflection as a goal for teacher education that there is a need to shift from merely individual teacher reflection to "emphasis on reflection as a social practice that takes place within communities of teachers who support and sustain each other's growth" (p.72). This is because challenging questions asked during the group discussion are vital to help teachers clarify their doubts and misconception while encouraging words and moral support will help teachers gaining confidence to pursue their goals. In this study, we attempted to encourage a group of in-service teachers to make reflection as a social practice during a workshop.

2.3 *Possible methods used for reflective thinking*

A literature search shows that there are various ways to facilitate reflective thinking. For example, Richards (1991) pointed out that journal writing can be used to: (a) provide a record of the significant learning experiences that have taken place; (b) help the participant come into touch and keep in touch with the self-development process that is taking place for them; (c) provide the participants with an opportunity to express, in a personal and dynamic way, their self-development; (d) foster a creative interaction (i) between the participant and the self-development process that is taking place, (ii) between the participant and other participants who are also in the process of self-development, and (iii) between the participant and the facilitator whose role it is to foster such development.

Another way to facilitate reflective thinking is through self-reports (Pak, 1985) which involves completing an inventory or check list of teaching practices, individually or in group sessions. Richards (1990) found that the accuracy of self-reports increased when teachers focused on the teaching of specific skills in a particular classroom context and when the self-report instrument was constructed to reflect a wide range of potential teaching practices and behaviours. In addition, self-reports allow teachers to make a regular assessment of what they are doing in the classroom by checking to see to what extent their assumptions about their own teaching are reflected in their actual teaching practices. For instance, self-reports could be used by a teacher to find out the kinds of teaching activities being regularly employed, whether all of the lesson objectives are being addressed, the extent to which personal goals for a class are being met, and the kinds of activities which seem to work well or not to work well (Richards, 1991).

Direct observation followed by immediate reflection is also commonly used to encourage reflective thinking. This is particularly common in teacher professional development models such as Lesson Study established in Japan since the 1960s. Lesson Study is a process by which small groups of teachers meet to collaboratively plan lessons, observe these lessons unfold in classrooms, discuss their observations and revise the lesson plans. It comprises six main steps: (1)

collaboratively planning the lesson plan, (2) seeing the lesson plan in action, (3) discussing the lesson plan, (4) revising the lesson plan, (5) teaching the new version of the lesson, and (6) sharing reflections about the new version of the lesson (Fernandez & Yoshida, 2004). A number of studies have shown that Lesson Study improves teachers' learning and supports teachers to grow professionally (Stigler & Hiebert, 1997, 1999; Shimahara, 1998; Lewis & Tsuchida, 1998; Yoshida, 1999; Lewis, 2000; Fernandez & Yoshida, 2004; Lim, White, & Chiew, 2005; Chiew, 2009). In addition, Fernandez and Robinson (2006) identified three main categories as central to pre-service teachers' learning through Lesson Study, namely connecting theory and practice, collaboration, and reflection.

With the advancement and availability of the digital tools, digital video is now an increasingly popular tool to facilitate reflective thinking. Digital video allows for the complexities of the classroom to be brought into sharp focus (Perry & Talley, 2001) by capturing the immediacy of the classroom and offering detailed and rich data on the teaching and learning process (Newhouse, Lane, & Brown, 2007). Capturing the voice, the behaviour and the movement of teachers and pupils in a powerful medium (Tochon, 2001) enables teacher trainees to view examples of authentic learning experiences. In fact, several studies (Harford & MacRuairc, 2008; Newhouse, Lane, & Brown, 2007; Rich & Hannafin, 2009; Rosaen, Lundeberg, Cooper, Fritzen, & Terpstra, 2008) have shown that digital video is a powerful means of facilitating critical reflection and furthering self-evaluation.

Apart from the rich information obtained through videoing and analysing one's own work, peer-video analysis (a process by which pairs of teacher trainees work collaboratively videoing and then viewing one another's work) can also promote dialogue and shared learning. When video reflection is scaffolded with guiding prompts, some studies (see Crawford & Patterson, 2004; Calandra, Brantley-Dias, & Dias, 2006; Brantley-Dias, Calandara, & Fox, 2007) indicated that beginning teachers can look beyond superficial commentary and write robust reflections. Crawford and Patterson (2004) examined how seven teacher trainees' reflections can be enhanced using video footage of their teaching and when the reflection process was scaffolded. They found that guided video reflections as well as timely feedback from their supervisors increased the depth of the teacher trainees' reflections.

Van Es and Sherin (2002) and Sherin and van Es (2005) examined how digital video could be used to help pre-service and in-service teachers in learning to notice what was happening in their classrooms. In their first study (Van Es & Sherin, 2002), six math and science teacher trainees reviewed video footage from their own and others' teaching and annotated it using a series of textual scaffolds embedded within a video analysis software called Video Analysis Support Tool (VAST). They observed that the teacher trainees' discussion changed from evaluative remarks like whether the activity went well or not to interpretive comments such as why this might have happened. In their second study, four middle school mathematics teachers met monthly in a video club to discuss excerpts from videos of their own classrooms and the discussions were facilitated by the researchers. Sherin and van Es (2005) found that the focus of the teachers' discussions shifted from pedagogy to student learning over the course of 10 monthly video club meetings.

Likewise, Yerrick, Ross, and Molebash (2005) employed digital video as a tool to foster beginning science teachers' reflections. They utilized an instructional approach that encouraged their participants to explore and reflect while they were doing their planning and teaching and while editing their video accounts as well. Yerrick *et al.* (2005) observed that there were shifts in the participants' reflections regarding planning and instruction, notions of teaching expertise, and pedagogical knowledge.

Similarly, Calandra, Brantley-Dias, and Dias (2006) explored how an urban, middle school science teacher trainee reflecting on her own teaching using an edited digital video. Her level of reflection was evaluated using the Framework for Reflective Thinking (Sparks-Langer, *et al.*, 1990). Initially, her level of reflective thinking was rather low when the reflection was unguided. However, she showed a significantly higher level of reflective thinking in the final stimulated recall interview session in which she linked theory to practice. Based on these findings, Brantley-Dias, Dias, Firsch, and Rushton (2008) developed a Critical Incident Reflection protocol to guide pre-service teachers' analysis of videotaped lessons. Their over-arching goal was to understand how using digital video and critical incident analysis might serve as productive tools in pre-service teachers' learning to teach process. They found that although digital video and the scaffolding support provided by the Critical Incident Reflection tool seemed to be beneficial in assisting the pre-service teachers to identify their strengths and weaknesses, most did not exhibit depth in their written analysis nor in their reflective discourse with the researchers.

The above reviewed studies demonstrated that combining digital video with scaffolds for reflection such as reflection guidelines, guided conversations and even digital video editing has the potential to help pre-service teachers notice classroom events and promote meaningful reflection albeit with some limitations.

3 The Study

The above review of related literature shows that reflective thinking is important and beneficial for pre-service teachers. In this chapter we address the under researched area related to reflective thinking involving in-service teachers. In our study we have attempted to promote reflective thinking among in-service teachers through lesson observation and the use of digital video recording. In brief, the main objective of the study was to promote reflective thinking among in-service teachers using a video recorded mathematics lesson so as to improve their teaching practices.

3.1 *The participants*

The participants of the study consisted of two groups of mathematics teachers, 18 primary mathematics teachers from two Chinese primary schools in Malaysia who attended a workshop conducted by the authors in Malaysia and another 39 Singaporean mathematics teachers who participated in a workshop conducted by the first author in Singapore.

3.2 *Data collection*

During the workshops, the participating teachers were shown a 10-minute video recorded mathematics lesson and asked to write their reflections about the lesson on a worksheet. The video recorded lesson was taught by a Malaysian primary mathematics Excellent Teacher. Excellent Teacher is an award given by the Malaysian Ministry of Education to teachers based on their expertise in subject matters and excellence in teaching.

To scaffold the reflective thinking, we provided the participants a worksheet (see Appendix A) comprised of six questions. For the purpose of this chapter we focus on two main questions: (1) "What do you think about this lesson?" and (2) "Given a chance, would you teach the lesson differently? Why? What materials would you choose to teach the lesson?" The participants were given sufficient time to reflect and write down their thinking. They were also allowed to discuss with the other participants as this was not an examination. The participants were also given the freedom to express their thought in whatever language that was comfortable to them. The data collected from the Malaysian Chinese primary school teachers were mainly written in Mandarin, those collected from the Singaporean primary school teachers were all in English.

The edited video-recorded lesson. The teacher who taught the video recorded mathematics lesson is a Malaysian primary mathematics Excellent Teacher. She is female, aged 55 years old, and has been teaching mathematics in Chinese primary schools for more than 33 years. The learning objectives of the lesson were to enable pupils: (a) to measure, read, and record masses of object in kilogram and gram using the weighing scales; and (b) to determine how the mass of an object compared to another. The teaching aids used by the teacher were weighing scales, packet foods, and worksheets. The sequence of the teaching and learning activities is presented as follows:

- First, the teacher displayed packet foods on the table. She asked the students what characteristics (types of food, price, and quantity) they would consider when shopping. Next, she facilitated a discussion with students about the mass of objects in daily life.

- Second, students were divided into six groups. Each group was given 5 types of packet foods. They were asked to measure, read, and record the mass of packet foods on the worksheet.

- Third, each group of students was asked to present their worksheets on the blackboard. The teacher compared and checked their answers.

- Fourth, the teacher asked the students to compare the masses of two packet foods by asking the following questions: "Which packet food is heavier?", "How many times is the mass of sugar as compared to the mass of rice?" and "What is the fraction of the mass of red beans as compared to the mass of green peas?".

- Fifth, the teacher encouraged the students to compare and record the mass of any of the two packet foods. Then the students presented their answers on the blackboard. This step assessed the students' understanding of the concept of comparative mass and corrected their misconceptions as well.

- Lastly, the teacher explained the strategies of finding the correct answers in detail so as to strengthen and enrich the students' skills of comparing masses of objects.

3.3 *Data analysis and discussion*

Responses to the first question: "What do you think about this lesson?"

Analysis of their responses to the first question showed similar reflections such as the lesson was lively and pupils were happily engaged in the classroom activities. About half of the participants (28 participants) reflected that the lesson allowed hands-on and active engagement of pupils as evidenced by the following excerpts: "The pupils were given a chance to weigh things and discuss" (S11); "engaging with lots of hands-on activities" (S30); "学生的参与度很大" [student engage a lot] (MB02); "每个学生有机会参与。自己秤，自己记录，可以加强对重量的认识" [every pupil has a chance to participate. Self weigh and self record, can enhance their knowledge about mass]. (MC05).

Fifteen participants pointed out that the teacher used real-life examples and encouraged collaborative learning through group work. They observed that the teacher "use examples found in their daily lives" (S13); and "Used real objects to weigh" (S11). They believed that, "Pupils work in group, they get to learn from one another" (S18) and also "可以组别活动中学习分工合作" [can learn to be cooperative from the group works] (MC06).

As a result, five participants felt that the lesson created an exciting atmosphere for the students to learn. Two Malaysian teachers wrote that: "学生乐在学习中" [students learning happily] (MC02) and "教学非常活泼生动， 很引导学生" [The teaching was very lively and attractive] (MB04). Likewise the Singaporean teachers also commented that "Learning math is fun - experiential learning (actions to seen)" (S25); and "can see children enjoy the lesson" (S24).

In addition to the above positive comments about the observed lesson, a few participants were concerned that some "Pupils do not know how to read scales" (S23); particularly the weaker pupils such that "学习缓慢的学生也许不会用秤来秤东西" [the slow learners might not know how to use the scale to weigh things] (MC09). Others commented that "秤的摆放会让坐在后面的学生看不太清楚" [the way the weighing scale was positioned might not allow the pupils who were sitting behind the class to see it clearly] (MB05). Consequently, some participants suggested the possibility of a "Buddy-system: pairing of students so weaker students do not fall behind" (S37) and also "Not all pupils are visual. Every explanation should have been written on the board for weaker pupils" (S10).

Although the above responses show consensus among participants about the observed lesson, we also noticed that the lesson has provoked the participants to reflect on many themes such as the overall teaching processes, students' reactions, the use of teaching materials, and particular focus on less successful pupils.

Responses to the second question: "Given a chance, would you teach the lesson differently?" "What materials would you choose to teach the lesson?"

Analysis of the second question shows that this open question provided more space and opportunities for the participants to reflect on themselves. If they were given a chance to teach the lesson, the majority of them would opt to teach the lesson differently. An interesting observation was that out of the 18 Malaysian participants, 8 of them agreed fully with the teaching method used by the teacher in the video. However, all Singaporean participants suggested otherwise. Perhaps due to cultural differences, Malaysian teachers may share certain similar characteristics or beliefs about mathematics teaching methods while Singaporean teachers may look at it from a different cultural perspective. Indirectly, this could imply that observing teaching from a different culture might encourage a deeper and better reflection as the observers are using a different cultural lens.

Numerous themes emerged from the responses as follows:

a) Group size

Seven Singaporean participants commented on the group size but this was not a concern to any of the Malaysian participants. S27 commented that, "Get pupils to work in pairs instead of in groups. If groups are too big, certain pupils tend to dominate the activity". This suggestion was agreed by S06 who noted that she would get "Pupils to work in pairs or threes". Likewise, S05 argued that, in the "second activity: Mostly class based so only a few pupils get to try. Perhaps can get them to do think-pair-share, then all pupils would have had a chance to try the questions".

It appears that Singaporean participants concerns about pupils' equity were expressed in the need for smaller group size or in pairs so that every pupil has the opportunity to learn. In addition, they suggested alternative strategies such as 'think-pair-share' to be integrated in group activity so that all pupils have equal opportunity to work on the questions or exercises given by the teacher.

b) Use of mini-whiteboard

In relation to the discussion on equity of learning, three Singaporean participants specified the use of mini-whiteboard. S17 remarked that "In the second part of the activity, I would get pupils to write their answers (to practise the questions) on individual small whiteboards so that everyone has a chance to practise the question". Similarly, S19 also agreed that, "I would get the whole class to participate by giving their responses on mini-whiteboards, red/green cards, wrong/disagree, right/agree." Their idea was also supported by S10 that "could have use a mini-white board for everyone to ensure all have learnt by writing their answers on the board".

c) Structured worksheet

Besides the use of mini-whiteboard, five Singaporean participants also proposed that the teacher provide structured worksheet or "mini-worksheets" (S38). This is because providing "worksheet so students don't need to spell" (S37). S39 explained that, "I would provide worksheets for the activity so they [the pupils] won't spend time trying to find/spell the items since this activity is focusing on finding mass. Therefore they should spend more time learning how to read the reading scale." The above comments indicate that these Singaporean participants were very much concern about allowing time for individual thinking and also promoting accountability among the pupils.

d) Use of weighing instruments

Some participants, that is two Malaysian and three Singaporean teachers were concerned about the use of weighing instruments. S10 proposed that she will "start by introducing different weighing scales. Teach the pupils how to obtain the sub-unit of each scale" and her idea was supported by S15, "Maybe good to teach how to read the scales first before the activity". Meanwhile, the two Malaysian participants suggested alternative weighing instruments such as "用天平来比较2种实物" [use balance to compare two concrete objects] (MC01) or electronic weighing machine (MB08) while a Singaporean participant suggested to "Use online manipulatives NLVM (online weighing machine)" (S16).

e) Giving alternative teaching ideas

Apart from commenting on the group size and teaching materials, reflecting on the observed lesson has generated many alternative teaching ideas. There were 22 responses that offered alternative teaching ideas or refinement of the activities such as the following:

A few participants proposed the importance of 'estimation' such that "I would have made the pupils estimate the mass of the product before actual weighing" (S14); or "Ask the pupils to estimate first before weighing" (S5). Another participant, S21 also stressed that "estimate first, then measure. Use smaller items first, estimate an apple's mass, then measure, move on to heavier item".

Other participants suggested having "learning stations". For example, S07 described that, "We do learning stations in our school for this type of activity. Five or six different stations with different items and weighing scales". Similarly, S13 further elaborated that, "Yes. I would probably provide the materials to weigh, and place them in stations. I'll have the students to walk around to each station to weigh the items. After comparing the answers as a class, I'll have the students to make the comparisons between the items".

Another alternative teaching idea was asking pupils to compare the printed mass on the packaging with the actual mass. S34, for instance, suggested that, "Using same items for each group - cover the mass on packaging (get the pupil to weigh and record). Then remove the tape and discuss the difference in masses on packaging versus measured mass". S22 also agreed that, "Mass of items was printed on package. May want to cover them or ask them to compare the masses with what they had weighed (to verify)". Or even "Repack the quantity! Get pupils to re-do and set their own questions" as suggested by S30.

Other teaching ideas suggested by the participants are as follows:

Activity 1: Discuss the difference between the interval scales in the 2kg and 10kg scales. Discuss why the student wrote 1.3kg and 1220kg. (S37)

Activity 2: In groups, get pupils to use the concept learnt to come up with mini-worksheets for other groups to work on. Groups to mark each other's work and explain each other's mistakes through presentation to whole class. Builds metacognition and confidence. (S38)

Activity on mixing syrup with water to make fruit punch to teach ratio/or in the context of cake-baking (proportion of sugar & flour). (S24)

Using significant episodes of the lesson to generate further discussion

During the workshop, after observing the video recorded lesson, the participants were given time to write down their reflections on the worksheet. After that, they were encouraged to share their reflections in pairs (with their neighbour), then in small groups and subsequently in a whole group. In all discussions critical incidents in the lesson brought out for attention by one or two participants initiated heated discussion which in turn, generated many more alternative ideas.

For instance, in the observed lesson different groups of pupils were given different weighing scales to measure the same package of milk powder. Two pupils from two different groups were then asked to write down the mass of the milk powder package on the board. Figure 1 displays what they had written. One wrote as "1.3kg" while another wrote as "1.220kg". The teacher then discussed with her pupils the possible reasons for the same package of milk powder to have different masses. However, the participants who observed the lesson argued that the teacher did not fully make use of the error in the discussion with her pupils. Some participants suggested that the teacher should allow the two pupils to re-measure the mass of the milk powder. Others suggested that pupils could be asked to compare the printed mass with the mass that they have measured. If the two values were different, what was the possible reason for this difference? Another participant highlighted that the printed mass is possible to be a bit different from the measured value as the printed mass is the "net weight". This then led to the discussion on what does it mean by "net weight" and does the teacher need to introduce the concept of "net weight" to pupils at this age?

Therefore, we found that the video recorded lesson could be used to initiate further discussions and reflections of significant episodes of the lesson among the teachers—discussions which could enhance teachers' understanding of the mathematical ideas and help them to generate new teaching ideas.

Figure 1. Two different masses of the milk powder package

4 Implications and Conclusion

Although journal writing and self-reports are useful tools for facilitating reflective thinking, they cannot capture the moment-to-moment processes of teaching and learning as many things happen simultaneously in a classroom. Besides, some significant aspects of a lesson may not have been observed by the teacher, let alone remembered, hence the value of digital video recording of the lesson to facilitate reflective thinking among teachers (Richards, 1991). In our study, using video recorded lessons for promoting reflection and as a professional development programme provided a potential platform for teachers to reflect on their own teaching while learning from others' teaching. The digital video recording of a lesson provided a convenient way of initiating reflection as it enables instant and repeated rewind of the lesson which in turn allows the teachers to revisit any incident, interaction and episode of action easily.

In sum, making reflections, whether reflection-in-action or reflection-on-action, can expose teachers to a world of new experiences and generate discussion of alternative teaching strategies and ideas. Observing other teachers teaching, whether in live or in video, provides a window for the teachers to look into how other teachers taught their lesson, and also how the pupils reacted to their teacher's teaching. In retrospection, the teacher can then reflect on oneself, and either decides whether to adopt or adapt some changes, or to continue with what they are doing now. Either way, engaging oneself in reflection is surely an experience that will be rewarding and enriching for the teachers which in turn can be a powerful impetus for improving their teaching practices. In particular, as shown in this study, looking at classroom teaching that is different from one's cultural setting is a powerful means of challenging one's thinking about effective teaching. Moreover, observing the video recorded lesson together with support groups help to scaffold discussion and reflective thinking.

References

Brantley-Dias, L., Calandra, B., & Fox, D.L. (2007, April). *Teacher candidates' experiences with digital video editing for reflection: How much scaffolding do they need?* Paper presented at the American Educational Research Association conference, Chicago, IL.

Brantley-Dias, L., Dias, M., Frisch, J.K., & Rushton, G. (2008, March). *The role of digital video and critical incident analysis in learning to teach science.* Paper presented at the American Educational Research Association Annual Meeting, New York City, New York.

Calandra, B., Brantley-Dias, L., & Dias, M. (2006). Using digital video for professional development in urban schools: A pre-service teacher's experience with reflection. *Journal of Computing in Teacher Education, 22*(4).

Chiew, C.M. (2009). *Implementation of lesson study as an innovative professional development model among mathematics teachers.* Unpublished PhD thesis, Universiti Sains Malaysia.

Crawford, B.A., & Patterson, B. (2004). *Scaffolding videotape reflections to enhance pre-service teachers' practice.* Paper presented at the 2004 International Conference of the Association for the Education of Teachers in Science, Nashville, TN.

Dewey, J. (1933). *How we think: A restatement of the relation of reflective thinking to the educative process.* Boston: DC Heath and Company.

Fernandez, M.L., & Robinson, M. (2006). Prospective teachers' perspectives on microteaching lesson study. *Education, 127*(2), 203-215.

Fernandez, C., & Yoshida, M. (2004). *Lesson Study: A Japanese approach to improving mathematics teaching and learning*. Mahwah, New Jersey: Lawrence Erlbaum Associates.

Harford, J., & MacRuairc, G. (2008). Engaging student teachers in meaningful reflective practice. *Teaching and Teacher Education, 24*(7), 1884-1892.

Harford, J., MacRuairc, G., & McCartan, D. (2010). 'Lights, camera, reflection': Using peer video to promote reflective dialogue among student teachers. *Teacher Development, 14*(1), 57-68.

Jay, J.K., & Johnson, K.L. (2002). Capturing complexity: A typology of reflective practice for teacher education. *Teaching and Teacher Education, 18*, 73-85.

Lewis, C. (April 2000). *Lesson Study: The core of Japanese professional development.* Paper presented at the Annual Meeting of the American Educational Research Association, New Orleans, LA. Retrieved May 17, 2011, from http://www.lessonresearch.net/aera2000.pdf

Lewis, C., & Tsuchida, I. (1998). A lesson is like a swiftly flowing river: Research lessons and the improvement of Japanese education. *American Educator*, 14-17 & 50-52.

Lim, C.S., White, A.L., & Chiew, C.M. (2005). Promoting mathematics teacher collaboration through lesson study: What can we learn from two countries' experience. In A. Rogerson (Ed.), *Proceedings of the 8th International Conference of the Mathematics Education into the 21st Century Project: "Reform, Revolution and Paradigm Shifts in Mathematics Education"* (pp. 135-139). Johor Bahru: Universiti Teknologi Malaysia.

Newhouse, C.P., Lane, J., & Brown, C. (2007). Reflecting on teaching practices using digital video representation in teacher education. *Australian Journal of Teacher Education 32*(3), 1-12.

Pak, J. (1985). *Find out how you teach.* Adelaide, Australia: National Curriculum Resource Centre.

Perry, G., & Talley, S. (2001). Online video case studies and teacher education. *Journal of Computing in Teacher Education, 17*(4), 26-31.

Rich, P., & Hannafin, M. (2009). Video annotation tools. *Journal of Teacher Education, 60*(1), 52-67.

Richards, J.C. (1990). *The language teaching matrix.* New York: Cambridge University Press.

Richards, J.C. (1991). Towards reflective teaching. *The Teacher Trainer, 5*(3), pp. 4-8.

Rosaen, C., Lundeberg, M., Cooper, M., Fritzen, A., & Terpstra, M. (2008). Noticing noticing: How does investigation of video records change how teachers reflect on their experiences? *Journal of Teacher Education, 59*(4), 347-60.

Schön, D. (1983). *The reflective practitioner: How professionals think in action.* New York: Basic Books.

Sherin, M.G., & van Es, E.A. (2005). Using video to support teachers' ability to notice classroom interactions. *Journal of Technology and Teacher Education, 13*(3), 475-91.

Shimahara, N. K. (1998). The Japanese model of professional development: Teaching as craft. *Teaching & Teacher Education, 14*(5), 451-462.

Sparks-Langer, G.M., Simmons, G.M., Pasch, J.M., Colton, A., & Starko, A. (1990). Reflective pedagogical thinking: How can we promote it and measure it? *Journal of Teacher Education, 41*, 23-32.

Stigler, J.W., & Hiebert, J. (1997). Understanding and improving classroom mathematics instruction: An overview of the TIMSS video study. *Phi Delta Kappan, 79*(1), 14-21.

Stigler, J.W., & Hiebert, J. (1999). *The teaching gap: Best ideas from the world's teachers for improving education in the classroom.* NewYork: The Free Press.

Tochon, F.V. (2001). 'Education-research': New avenues for digital video pedagogy and feedback in teacher education. *International Journal of Applied Semiotics, 2*(1-2), 9-28.

Tremmel, R. (1993). Zen and the art of reflective practice in teacher education. *Harvard Educational Review, 63*(4), 434-458.

Valli, L. (1997). Listening to other voices: A description of teacher reflection in the United States. *Peabody Journal of Education, 72*(1), 67-88.

Van Es, E.A., & Sherin, M.G. (2002). Learning to notice: Scaffolding new teachers' interpretations of classroom interactions. *Journal of Technology and Teacher Education, 10*(4), 571-596.

Yerrick, R., Ross, D., & Molebash, P. (2005). Too close for comfort: Real-time science teaching reflections via digital video editing. *Journal of Science Teacher Education, 16*, 351-375.

Yoshida, M. (1999). *Lesson Study (Jugyokenkyu) in elementary school mathematics in Japan: A case study.* Paper presented at the American Educational Research Association (1999 Annual Meeting), Montreal, Canada.

Zeichner, K., & Liu, K.Y. (2010). A critical analysis of reflection as a goal for teacher education. In N. Lyons (Ed.), *Handbook of reflection and reflective inquiry* (pp. 67-84). USA: Springer.

Appendix

Name/ Nama/姓名: ______________________________

School name/ Nama sekolah/校名: ______________________________

Questions for discussion/ Soalan untuk dibincang/讨论的问题:

1.What do you think about this lesson?
Apakah pendapat anda tentang pelajaran ini?
你对这节课有什么想法?

2.What can the students learn from this lesson?
Apakah perkara yang boleh dipelajari oleh murid-murid dalam pelajaran ini?
学生可以在这堂课里学到什么?

3.Which activities engage the students? Please give some examples.
Activiti yang mana melibatkan murid-murid? Sila beri contoh.
有那些活动让学生参与? 请给予一些例子。

---

4. Do you think the teacher's questions stimulate students' thinking? Give some examples.
Adakah soalan guru merangsangkan fikiran murid-murid? Sila beri contoh.
老师给予的问题可否激励学生的思维? 请给予一些例子。

---

5. Do you think the teacher had achieved the lesson objectives? Why?
Adakah guru telah mencapai objektif pelajaran? Mengapa?
老师是否已经达到这堂课的目标? 为什么?

---

6. Given a chance, do you teach the lesson differently? What materials would you choose to teach the lesson?
Jika diberi peluang, adakah anda akan mengajar pelajaran ini secara berbeza? Apakah bahan-bahan yang anda akan pilih untuk mengajar pelajaran ini?
若有机会，你是否会运用不同的方式来教导这堂课?
你会用什么教材?

Chapter 7

Learning from Student Reflections

Barry KISSANE

Nurturing reflective learning in the school depends critically on teachers being appropriately reflective themselves. In this chapter, the author describes some pre-service primary teachers reflecting on their own experiences of learning mathematics. The pre-service teachers were required to reflect on how they had learned mathematics and on how technology was used to support their learning. The chapter draws on a sample of the resulting reflections to illustrate the value of such an activity for the pre-service teachers themselves, while the results of the reflections also provided learning opportunities for their teacher.

1 Introduction

Good teachers are often described as 'reflective practitioners', suggesting that reflection is a characteristic of sound professional practice. This chapter explores some benefits of engaging in reflective thinking about mathematics, from a teacher education perspective.

The choice of such a perspective should not limit the scope of the chapter to teacher education, however. In the first place, teachers are generally expected to engage in personal professional development, so that thinking about the reflective process does not end on graduation from teacher education, but continues long into professional practice. In the second place, mathematics teachers often strive to encourage their students to think deeply about what they are learning, in order to help them make connections between aspects of mathematics for themselves.

The recent interest in metacognition among both teachers and researchers is symptomatic of a widespread view of the merits of encouraging students to think about their thinking, at all levels of education.

So this chapter describes some experiences that were designed to encourage pre-service teachers (the students) to reflect on aspects of their learning and considers what their teacher (the teacher educator) might learn from this also, recognizing that teachers are also learners themselves.

2 Review of the Literature

The twin ideas of metacognition and reflective thinking have been discussed in educational circles for many years now, although it is still unusual for them to be explicitly highlighted in school curricula, as they are in the framework for school mathematics in Singapore (Ministry of Education, 2006). Lesh and Zawojewskii (2007, p. 770) noted Flavell's seminal introduction of the term 'metacognition' in 1976:

> Metacognition refers, among other things, to the active monitoring and consequent regulation and orchestration of these processes in relation to the cognitive objects or data on which they bear, usually in the service of some concrete goal or objective. (p. 232)

They also noted the variety of ways in which the term has been interpreted by researchers and practitioners alike, so that a uniform meaning has not developed. To illustrate this, Wiliam (2007, p. 1079) suggested that, "It is perhaps not too much of an overstatement to say that everyone agrees that metacognition is important, but no one can agree on what it is, at least in a sufficiently precise definition to put into practice". Possibly for this reason, many interpretations of the term have used the unsophisticated, but helpful, idea that it is concerned with 'thinking about thinking'.

The importance of reflective thinking for education was recognised many years ago by the American educational philosopher John Dewey, as noted by McDonald and Dominguez (2009):

> Dewey (1933) suggests that reflective thinking is an active, persistent and careful consideration of a belief or supposed form of knowledge and the further consideration to which that knowledge leads. Learners are aware of and control their learning by actively participating in reflective thinking–assessing what they know, what they need to know, and how they bridge that gap–during learning situations. (p. 46)

After reviewing the case for students to write reflectively in science, McDonald and Dominguez (2009, p. 49) summarised the argument by suggesting, "In short, reflection creates meaning".

The Singapore Primary Mathematics syllabus (Ministry of Education, 2006) highlights metacognition as one of five key components of it's framework. Interestingly, however, most of the detailed advice in the syllabus document refers to the importance of metacognition in the process of problem-solving (Ministry of Education, 2006, p. 9). In order to understand its significance in student learning, Wilson (1998) developed new approaches for accessing metacognition in students, focusing on its use in solving mathematics problems. She suggested that a variety of procedures ought be used in what she described as a 'multi-method interview', based on various kinds of observations of students addressing a mathematical problem. While problem solving seems likely to be helped by metacognition, attention is also due to the notions of metacognition and reflective thinking in the process of *learning* mathematics, as well as the place of these ideas in thinking about the nature of mathematics itself.

When considering claims that reflective thinking can be a positive element in mathematics learning, it is important to consider carefully what constitutes mathematics learning. For example, Wheatley (1992) highlighted potentially fruitful relationships between reflective thinking and learning mathematics, presenting evidence that "...encouraging reflection results in greater mathematics achievement, even on standardised tests which stress procedures and particular conventions". (p. 529). He distinguishes the broad approach of 'problem-centered learning' from two other broad approaches to mathematics in school, which are still distinguishable today. One of these is the "explain-practice" approach, a standard feature of traditional school mathematics, focusing on making sure that students develop the procedures regarded as appropriate for school mathematics as a first priority. These may indeed be the procedures that are regarded as necessary or especially important for formal assessment. A second approach is what Wheatley describes as "abstract-first". He notes that

> The widespread use of the term "application" suggests that mathematics is viewed by many as a formal system to be learned in the abstract and then "applied to the real world". (p. 533)

These approaches might even underpin what would otherwise be seen as non-traditional approaches to teaching and learning, as Wheatley further notes in discussing the use of manipulatives for learning mathematics. He argues that using concrete objects to demonstrate mathematical ideas is still based on an 'abstract-first' approach, however. Thus, referring to the use of base ten blocks, which are commonly used by teachers to represent decimal place value, he notes:

> Often when manipulative materials are used in teaching mathematics, the teacher demonstrates *the* way they are to be used and students are left little freedom to give meaning to the experience in ways that make sense to them; the way the materials are used is prescribed. There is the mistaken belief on the part of the teacher that the mathematics is apparent in the materials. ... This is based on the belief that the mathematics is "out there" and that models "show" the concepts. The demonstration with concrete materials is quite appealing because the concepts are so vivid for those who have *already* made the construction. Thus there is the mistaken belief that since we, as adults, can see the mathematics in the blocks, the students will too. But the "seeing" requires the very construction the activity is intended to teach. (1992, p. 534)

Although virtual manipulatives were developed and used by teachers long before widespread Internet access and use, they were not as prominent at the time of Wheatley's paper as they are today. However, similar scrutiny of arguments might be made regarding the use of these as well as concrete manipulatives. In short, the nature of students' learning depends on what they do with the materials, not on the materials themselves. Wheatley concluded that the prime focus of teaching mathematics is to help students to make sense of it; he provides evidence of the potential for student reflection to contribute to that intention, noting that:

> In mathematics learning, the intention to make sense is essential. ... Neither the abstract-first nor procedures-first approach to learning fosters the intention to make sense. (1992, p. 533)

Such a concern for making sense of mathematics has deep roots, and arguably goes to the heart of the nature of mathematics. The very influential paper of Richard Skemp (1976) highlighted two different ways in which teachers and students seemed to routinely use the word 'understand' in relation to mathematics. He suggested that some interpreted 'understanding' as knowing what to do in a given circumstance, which may mean little more than having sufficient procedural knowledge to respond to routine tasks successfully (such as to use a formula to find the area of a triangle), and that such a colloquial interpretation was not uncommon; Skemp used the term 'instrumental' to describe limited understanding of that form.

In contrast, he highlighted the much greater power of what he described as 'relational' understanding, which would be exemplified by knowing *why* the area of a triangle is given by a particular formula. Relational understanding is what Skemp claimed he thought everyone meant by understanding: that it involves understanding how and why mathematical ideas are related to each other, not just 'understanding what to do'. Skemp (1976) noted that inconsistent interpretations of the idea of 'understanding' lead to many potential mismatches in school, such as mismatches between the intentions of students and teachers or mismatches between the curriculum and a textbook or an exam.

Ideas of sense-making and understanding are commonly highlighted in modern mathematics curriculum documents. Thus, the Singapore Primary Mathematics syllabus (Ministry of Education, 2006) notes:

> They should be given a variety of learning experiences to help them develop a deep understanding of mathematical concepts, and to make sense of various mathematical ideas, as well as their connections and applications, in order to participate actively in learning mathematics and to become more confident in exploring and applying mathematics. The use of manipulatives (concrete materials), practical work, and use of technological aids should be part of the learning experiences of the students. (p. 6)

Similarly, the Australian Curriculum: Mathematics emphasises understanding by identifying it as one of the four Proficiency Strands around which the curriculum is organized. The official curriculum (which is online) defines the Proficiency of Understanding as follows:

> Students build a robust knowledge of adaptable and transferable mathematical concepts. They make connections between related concepts and progressively apply the familiar to develop new ideas. They develop an understanding of the relationship between the 'why' and the 'how' of mathematics. Students build understanding when they connect related ideas, when they represent concepts in different ways, when they identify commonalities and differences between aspects of content, when they describe their thinking mathematically and when they interpret mathematical information. (Australian Curriculum, Assessment and Reporting Authority, 2012)

These kinds of official descriptions of the intentions of mathematics curricula and the nature of understanding seem to imply that student reflection is a necessary aspect of their learning, as they refer to mental operations which students are expected to undergo. The phrase 'deep understanding' in the Singapore syllabus might even be interpreted to refer to something beyond Skemp's idea of instrumental understanding, and akin to his idea of relational understanding.

In summary, research has suggested that there are potentially important links among reflection, metacognition and important learning of mathematics. This chapter explores some of these links in practice, considering also the role of the teacher, by considering some examples of reflections of pre-service primary teachers.

3 Two Assessment Tasks Involving Reflection

To encourage pre-service teachers to engage in reflective thinking, as well as to obtain information on the nature and quality of their learning, the author recently included activities involving reflection into a pre-service primary teacher education unit in Australia. One purpose for doing so was to encourage pre-service teachers to engage in reflection as part of their own learning, consistent with the overall theme of the first year of their teacher education course, which was 'Know yourself as a learner', as well as being concerned with an overarching goal of their course which is the development of 'reflective practitioners'. The pre-service teachers were provided with advice on undertaking personal reflections in another unit in their course, where they were expected to construct learning journals for that purpose.

The unit upon which this chapter was based was an elective for first year pre-service teachers whose mathematics background was weaker than required to be a primary school teacher, and so it focused on the development of mathematical content knowledge, rather than pedagogical content knowledge (which is addressed at later stages of their course). So an important purpose of the tasks was also to emphasise the potential value of being reflective in order to improve the likelihood of deep learning occurring. An additional, although longer-term, purpose was to begin a process of pre-service teachers appreciating the possible merits of reflective thinking by the (school) students they would eventually teach.

Two formal assessment tasks required the pre-service teachers to reflect on aspects of their own learning. These tasks provide some information regarding the reflections and also some insights for the teacher (i.e. the author in this case) regarding the pre-service teachers and their mathematics learning. They are described here in the expectation that both the substance of the reflections and the reflection process itself will be of interest to readers who are teachers, not only to those who are teacher educators.

3.1 *Reflections on learning*

One assessment task required the students (who were pre-service teachers) to reflect on their own learning, and to describe the learning process involved. The students were required to chose an aspect of mathematics that they had recently been engaged in learning, and to reflect on which learning activities were helpful to them, which were not helpful and why that was the case. Their work was to be based in part on informal reflective learning journals that they were encouraged to complete every week. Students had access to a range of resources for learning, including formal classes, textbooks, Internet sites and computer software but were left to their own devices to some extent to make choices among these.

3.2 *Reflections on learning with technology*

A second assessment task required students to reflect on their experiences of using technology, such as Internet resources, computer software or calculators to learn some mathematics. They were regularly referred to resources of these kinds, consistent with a view that teachers in modern times need to know about and use such resources in their teaching. In this case, the reflections were related to using the resources as learners, rather than as teachers, however. Students had a free choice of which particular piece of technology to use as a focus for their reflections, although popular choices involved the use of various virtual manipulatives such as those in the National Library of Virtual Manipulatives (Utah State University, 2010), *Interactivate* (Shodor Education Foundation, 2012) and *Illuminations* (National Council of Teachers of Mathematics, 2012). Another popular choice was a calculator, as all students were expected to own and use one.

4 Selected Themes Emerging

In this section, a number of themes that emerged from student reflections are identified and explored, with suitable exemplification provided from (anonymous) student writing. While some of these themes emerged in response to particular tasks, many student reflections gave rise to several themes. There is no intention here to systematically quantify the themes and their prevalence — that is to claim to represent these cohorts of students in a statistically faithful way. Nor should the order in which the themes are presented be interpreted as reflecting their relative frequencies of occurrence or their importance. Rather, the themes have been selected to highlight some of the potential outcomes of providing students with an opportunity to reflect on their own learning in these ways, and to indicate the kinds of issues that emerge. Indeed, the themes chosen for this purpose are regarded as those which are likely to be likely to be recognizable as relevant themes for teachers and student at various levels, and not only for pre-service teacher education.

The following selections also illustrate some ways in which teacher educators or teachers more generally might learn about their own students through the use of such tasks. The treatment of this issue is suggestive rather than comprehensive, in view of the available space. However, some themes have been chosen to illustrate how the teacher might learn from student reflections, in addition to (and, at times, instead of) students learning from student reflections.

4.1 *Personal histories and emotions*

Many pre-service teachers used the opportunities for reflection to speak frankly of their personal histories — frequently unfortunate — with mathematics in school. Some of these were especially graphic, such as the following two reflections, from two different students:

> *Maths, to me, has always stood for My Academic Torture Has Started. All of my life I have struggled with maths. I don't like it and I don't get it.*

> *Maths was my least favourite subject at school and is something that I still struggle to understand. No matter how hard I try, my mind just doesn't seem to retain the information and I can never see the relationships between the numbers and the words I'm hearing or reading. And, now when I hear the word "Maths" I instantly start to panic as I'm aware of my personal inadequacy in mathematics. As a result, I have learnt to avoid maths.*

Provided with an opportunity to do so, many students offered emotional responses of these kinds to mathematics: systematic avoidance, self-doubts, lack of confidence and many other negative emotions. For some students, it appeared that there was an element of catharsis attached to being permitted to express freely their relationships with mathematics, although the tasks did not explicitly require them to do so. Reflections of these kinds were informative for the teacher in understanding the students in the unit; sometimes they also revealed that the students were beginning to consider their role as teachers:

> *This topic has made me reflect on the way I see maths because I have always felt intimidated about maths and had the opinion that people were either good at maths or not. And I was one of those unfortunate people that were incapable of ever understanding maths! I don't feel this way any more, I believe that if people don't understand maths they just need it explained to them in a different way. The more I study the concepts the more I am starting to make connections and see how the concepts are intertwined and I feel like I am 'getting it'.*

Some pre-service teachers also showed reassuring signs of progress with their thinking:

> *Whilst explaining [this equation] and the fact that I used to have trouble doing this kind of calculation, I felt rather ridiculous. It seems such a simple mathematical problem to me now, however when I was at school, I was terrified of maths, all because I could not grasp it.*

Thankfully my mind and knowledge has matured with me and I don't feel so lost in a world that was once so unfamiliar to me.

Emotional responses to mathematics are well known to all teachers, so that it is unsurprising that students might use an opportunity to reflect on their learning to highlight these. Indeed, the unit textbook (Haylock, 2010) even has a chapter at the beginning devoted to a discussion of these sorts of responses. While reflecting on these affective elements was not the major intention of the tasks, it is hoped that the process of doing so has been advantageous for both the pre-service teachers and their teacher.

4.2 *Assessment*

Many pre-service teacher reflections referred to the key role of assessment in mathematics, incidentally encapsulated in student observations about their past practice with mathematics. For example, comments such as the following were not unusual, revealing both the primacy of formal testing as well perhaps as their conception of the nature of mathematics:

At best I've trained my mind to remember the right procedures for the coming test with no real understanding of the mathematical topic.

Some students, such as the following, observed that formal assessment in mathematics might even have a more significant role than it does in other aspects of the curriculum:

If someone asked me how to do a maths question from my Year 12 exam I would probably freeze on the spot and have no idea. But if they asked me a question from my history exam I would know the answer and be pretty confident to talk about it. Knowing this just re-instated my feelings of failure when it came to maths. I learned maths the ROTE way. I never committed anything to memory. I simply tried to get through life without actually learning how to do maths properly.

In studying to be a teacher, albeit a distant goal for these first year pre-service teachers, it is perhaps reassuring that reflections of these kinds provide evidence of a level of awareness of their mathematical thinking, problematic as it might be. In some cases, students used the process of reflection to reveal a gradual change of perspective, consistent with their growing realization that the kind of understanding needed for teaching mathematics needs to be more substantial than that expected to survive formal assessment processes. An example of this is the following:

I am no longer just trying to find the correct number to stick into a calculation to find the right number that will give me a tick. I am seeing and learning that all maths is interlinked and having strong fundamental building blocks is the key to understanding maths.

Indeed, it is perhaps also of interest that some pre-service teachers objected to the idea of being asked to reflect on their own learning in mathematics, because they had a very strong view, no doubt reinforced by years of experience, of what assessment in mathematics should comprise (which did not accommodate the idea of writing about their learning).

Frequent references to assessment may be a consequence of the status of the students as first years, for whom the most recent experience with mathematics was likely to have been at the end of secondary school, when external examinations cast a significant shadow over the school curriculum. It is valuable for all teachers to be aware of such strong feelings about the place of assessment in mathematics. We need to look for ways of reducing negative effects of assessment, while emphasising the positive effects of providing feedback on student progress.

4.3 *Rules of thumb*

Not all reflections provide positive evidence of student learning, although they may still be informative for the teacher. A number of student reflections made clear that their understanding of mathematics was what Skemp (1976) would describe as 'instrumental' in character or what Wheatley (1992) would describe as reminiscent of a 'procedures first' approach to mathematics. A typical example is the following:

> *A friend in my tutorial showed me the rule of what she called cross multiplying to find a common denominator. This is when you multiply each of the denominators by one another and then follow the rule that whatever you do to the bottom you do to the top. I was a little skeptical of this so I redid the equation* $^{2}/_{3} + {}^{3}/_{4} = ?$ *again but this time cross multiplying. I am really glad that I was introduced to this idea of cross multiplying as it will allow me to solve much larger sums of fractions.*

Rules of thumb of this kind abound in elementary mathematics, and provide students with convenient and efficient ways to undertake the kinds of tasks that they regard as necessary to demonstrate their progress with mathematics. There are many other examples, such as "invert and multiply" for division by fractions and "change the side and change the sign" for manipulations associated with solving equations.

In this case, the rule for finding the sum of two fractions with unequal denominators was clearly seen by the student as beneficial, although no evidence is offered that the student understood why the processes described were appropriate. In class work related to fractions, rules of this kind were not mentioned at all, in fact, but rather there was heavy emphasis on the key idea of equivalent fractions as a way of making sense of such a situation. While it is concerning that students seize upon such rules of thumb, in this case via a well-meaning colleague, it is nonetheless informative to the teacher to become aware of it. It is also concerning that the student writing this reflection seemed not to be aware of the limited understanding associated with the process used, but merely comfortable that it produces the correct answer, reminiscent of the concerns expressed by Skemp (1976).

4.4 *When are we gonna use this?*

A frequently recurring theme in student conversations and hence also in their reflections about mathematics relates to its fundamental purpose. Thus, many reflections include remarks such as those at the beginning of this pre-service teacher's comments:

> *Fractions, what are they? Where will I use them? Why do I need to learn about them? These were all questions I used to ask when I was at school. I never could understand what the teacher wanted us to do. I could never understand the concept of a common denominator or why we needed to times by the bottom number and then do the same for the top with the bottom number from the other fraction.*

It is never quite clear whether questions of these kinds are asked in earnest, to find out where mathematics is actually used, or whether they are symptoms of a different problem. Secondary school teachers commonly report that their students want to know why a particular aspect of mathematics is included in their study and that they usually are not satisfied with an observation that they will need it 'later'. Indeed, concern for such questions has lead to significant change in mathematics curricula in recent years, with much more focus on everyday applications than was the case a generation or two ago. A glance at modern textbooks will demonstrate the significant attention given to questions of the 'uses' of mathematics, presumably to appease the appetite of students to find out. Wheatley's (1992), observation that mathematics might be approached from an 'abstract first' perspective and then to seek 'applications' later is reminiscent of the same perspective.

Yet reflections of the kind reported here perhaps suggest another interpretation. Perhaps a request to find out where the mathematics is used is not in fact always a request for information at all, but rather a kind of *cri de coeur*, an (often impassioned) outburst triggered by confusion or a lack of understanding of key mathematical ideas. In the extract above, the student is both describing what they don't understand (the meaning of the concept of common denominator and the reasons for some particular procedures to be used) as well as indicating that they are puzzled about the purpose of the mathematics. (They do not, however, take a further step of reflecting on links between the two issues of use and understanding.) In support of this interpretation, it is revealing that very few students, if any, asked about the reasons for doing mathematics when they had a confident and rich understanding of it.

Although the reflections collected here were in respect of mathematics, it seems that students rarely ask about the usefulness of other aspects of the curriculum, or at least to the same extent. It is entirely plausible — albeit hypothetical — that our earnest endeavours over recent decades to make clearer where mathematical ideas are used may have legitimized the questions about the use of mathematics as part of an officially sanctioned way of thinking about mathematics, and we have not been as observant as we might have been about the underlying issue of understanding. It is possible that problems of this kind are particularly evident in the secondary school, and possibly emphasized in teacher education because many pre-service teachers have only recently left secondary school.

4.5 *The one right way*

A regular theme in pre-service teachers' reflections concerned the widespread view that, in mathematics, there is one 'correct' way of doing things, that he teacher knows what it is and that the students are expected to learn this and reproduce it on request. The following reflection is an example:

> *It was always explained as if there were only one way to do something, and if you were not able to do it their way you were ridiculed. I have always been quite a shy person, and not one to put my hand up or approach the teacher if I was having difficulty.*

For many students, such as the two following, such views of orthodoxy suggested that repetition was a powerful learning strategy:

Again to remember this type of method I spent time repeating similar [fraction] conversions so that my brain would retain it.

In maths, you need practice, practice and more practice.

It is encouraging to hear pre-service teachers beginning to question this orthodoxy a little, or at least to begin to listen to other opinions regarding it:

> The old algorithms are akin to learning by rote and they don't encourage real understanding of the mathematical concepts. ... An interesting point that I came across while studying the text is that there is no 'right' way to perform a calculation and that children, parents and grandparents should be made aware of this. (Haylock 2010, p. 99)

The idea that here might be several ways of thinking about things is very important for teachers, as students as a matter of course will think about things in different ways. Hence, it is important for pre-service teachers to begin to recognise that a preoccupation with the one right way of doing mathematics or thinking about a mathematical idea or operation is ultimately very limiting and not consistent with a relational understanding.

4.6 *Mental arithmetic*

In Australian curricula, each of written, mental and calculator methods of calculation are recognized as important for students and so are also important for teachers. Some emphasis was placed upon mental arithmetic in the unit, because it was felt that pre-service teachers tended to rely too much on standard methods of written computation, and even calculators, and were relatively inexperienced with mental calculation.

A number of students reflected on these experiences. For example, the following student reinforced our preconception that mental arithmetic could not be assumed amongst students, following some class work involving partitioning numbers mentally to facilitate addition:

The first time I observed partitioning was in the first maths lecture. This appeared to me to be a bizarre way to calculate a sum. In school (over 30 years ago), I was taught to begin calculations from the right hand side. I was taught to calculate vertically down the page, use the 'rules' to subtract, borrow 10 from the neighbour and put it on the doorstep. ... So this new way of calculating a sum horizontally seemed extremely strange to me. I could not relate to this horizontal approach at all. I felt extremely anxious in the lecture theatre; all the other students appeared to be answering the questions nodding their heads agreeing with the calculations. I was astonished. How could everyone calculate the answer mentally, without pen and paper?

Reassuringly, several students recognized that mental computation was often idiosyncratic, and required a good understanding of the underlying structure of numbers, unlike written computation, which was often completed without much thinking:

I think this is what I missed out on as a student. I went straight to do this, this and this and you get the answer this and that is right. I don't believe I ever truly understood the process behind what I was doing, nor was I aware of all the different ways one calculation can be done or one number can be seen. This tool of mental maths will be something that I will take into my teaching years.

Reflections that focused on mental computation were informative for teachers of these pre-service primary teachers, as they made clear the extent to which mental arithmetic seemed to have been neglected in their recent past. Perhaps because of the growing influence of machine computation in school and perhaps because of limited explicit attention to mental computation in the pre-service teachers' recent past in secondary school, the reflections emphasized the importance of explicitly targeting mental arithmetic expertise, rather than assuming that it would develop without attention.

4.7 *It's simple ...*

Perhaps especially when it is characterized as a collection of procedures, mathematics is often described by those who have mastered it as "simple" or "easy". Although such remarks are no doubt intended to be reassuring, it is rarely helpful for students to be informed by others that something that they are struggling to comprehend or master is simple, however. Reflections such as the following illustrate this issue:

> *So what is a factor again? I know I hated prime numbers and didn't understand them but multiples? They are just what you multiply together, right? My goodness, my heart rate started to rise, when the simple statement of "everyone remembers what a factor is, right?" was asked. I basically curled up in my chair in front of my computer and thought, "How can I consider myself reasonable at maths if I can't recall exactly what a factor is."*

The potential for students to be intimidated by what seem at first to be careless linguistic slips is substantial, if reflections of these kinds are considered carefully. There are lessons here for the teacher's own use of language and also lessons for attending to the prevailing language in the mathematics classroom. For example, students seem unlikely to be comfortable revealing their lack of understanding something if it has already been described as 'simple' or when there are public statements that everyone can understand it easily.

Learning new ideas is inevitably difficult, and involves students making errors, in the same way that solving difficult problems inevitably involves some element of confusion and misunderstandings along the way. Part of the role of the teacher is to help students recognize the inevitability of making errors and to use them productively, rather than allow them to be a source of undue pain. From the point of view of the pre-service teachers, reflections of these kinds might be used to alert them to be careful that they do not repeat with their own students in school, when they become teachers, the problems of this kind that hey encountered themselves in school.

4.8 *Visual learners*

Many pre-service teachers expressed a preference for 'visual' learning in their reflections. In doing so, many of them referred to their use of the online learning system *HOTmaths* (Cambridge University Press, 2011), which was used at times in formal classes and which they were also expected to make use of in their own time, as this included many examples that were visual in nature. For example, a typical student reflection noted:

> *I am mostly a visual learner, as I find it much easier to learn from diagrams and pictures like those in HOTmaths, than I do from words and symbols like those in the textbook. These just tend to confuse me.*

As Wheatley (1992) noted, however, and the next student echoed, students do not necessarily learn from visual images what was intended:

> *I remember my classes with fractions because the teacher gave us a sheet with fractions represented with circles, squares or rectangles that we needed to colour in according to the fraction. It was very easy, I used to enjoy it. Unfortunately, I paid more attention in my colouring in than to learn about fractions and their relations with each other.*

Visual images can be compelling, especially for those who are using them, although reflections of this kind suggest some caution in assuming that they offer the same advantages to the learner that they might to the person constructing them (such as a teacher). Although textbooks frequently include many images, and the use of interactive whiteboards seems to have increased the popularity of visual representations of mathematical ideas among publishers as well as teachers, care is needed to ensure that he meanings intended are the focus of the attention.

A good example of this issue concerns a visual proof of the Pythagorean Theorem, available in many printed materials and also on the Internet, such as on the *Cut-the-Knot* website (Bogomolny, 2012). In the author's experience, proofs without words of this kind are often described quite differently by those who can see the arguments involved than by those who do not see them at all, and are simply confused by them. It is questionable whether some students learn from images or find them informative unless they have already understood something, and it is not difficult for teachers to be in the same position.

4.9 *Virtual manipulatives*

Related to the previous theme, as the students were preparing to be teachers, some emphasis was placed on personal use of various forms of technology for learning, to begin familiarizing students with what is available in modern classrooms. It was clear that some students were initially reluctant to make much use of these, perhaps because they had been accustomed to learning mathematics in other ways. Indeed, it seemed that some students undertook the task of reflecting on their use of technology only because there was an assessment expectation that they do so; while of course this ought not be a major purpose of assessment, it is a useful ancillary effect. Despite such ambivalence, there were some students who were clearly enthusiastic about the change of learning experience, such as the following:

> *Technology today is fantastic and the graphics are outstanding. When I compare how I was taught maths in the UK in a cold damp classroom with a pen and paper, watching the teacher write on the blackboard in chalk, only being allowed to speak when answering a question or repeating times tables in the rote way of learning. Learning maths today is fun (I cannot imagine that I would ever say that!)*

Many pre-service teachers reflected on their use of virtual manipulatives in particular. For example, after using the *Area Explorer* applet in *Interactivate* (Shodor Education Foundation, 2012), which engages users in considering areas and perimeters of polygons simultaneously to help them distinguish these two concepts more clearly, one wrote:

> *I believe that even though it is important that children know how to get the 'correct answer' using a formula, it is more important that they really understand why we use the formula in the first place. Through exploring with this activity, I have been able to gain a clearer understanding of this.*

Similarly, after using the *Factor Tree* in the National Library of Virtual Manipulatives (Utah State University, 2010) to see how the product of the factors of a number equals the number itself, another pre-service teacher wrote:

> *This applet helped me to actually see and understand this relationship, which before I had not completely grasped. I knew ... what a factor and a prime number was but I hadn't made the connection between them until I used this applet. Using the Factor Tree helped reinforce my understanding of factors and primes building my confidence in using them. It allowed me to 'see' how the factors were linked and how to break them down.*

After reflecting on explorations with the applet to find greatest common factors and least common multiples, this student continued:

> *But what does knowing the GCF and LCM really tell me, what do the numbers mean? This is the interesting part of numbers and mathematics that I am learning for the first time during this unit. Previously I would have been content with the computer telling me that I was right and not thought about it further. Now I am interested in what that number means and what it is telling me.*

While these reflections are encouraging, it is also clear that students do not always learn well from experiences with manipulatives, as was noted earlier by Wheatley (1992) regarding concrete manipulatives. For example, the following pre-service teacher reflected on their use of a virtual manipulative concerned with probability simulation:

The student is expected to use the information given to determine a) how many sections the spinner has, and b) what colour each section should be. Once students have created their spinner they have to save it, then "test" their spinner. When testing their spinner students can select how many spins the spinner should do, and the speed. The application then shows the results in table and graph formats. The graph format provides a clear side by side comparison of the information, so student can compare their spinner with the original information.

The reflection continued in this vein to focus on how the virtual manipulative worked, but providing little inkling of what was learned from using it (such as the irregularity of results each time, with an overall recognisable consistency, leading to the idea of probability as long-term relative frequency). It is not clear from the account that what was learned was much more than how to operate the virtual manipulative successfully, when the real purpose was to help them learn mathematics.

4.10 *Calculators*

Reflections are not always as informative for the student as they are for the teacher, unfortunately. In reflecting on learning with technology, the pre-service teachers were able to choose any technology they wished, and a calculator was a popular choice. In classes, there was some stress on the potential for calculators to help students by seeing and exploring alternative representations of numbers (such as fractions and decimals). However, it was clear from student reflections that only rarely did students understand the ways in which they might learn mathematical ideas from using a calculator, as the focus was overwhelmingly on the calculator as an answering device. For example, one student wrote:

I used the four-function calculator all through primary school to verify the answers to questions I had done mentally to make sure I had got the correct answer. This helped me learn maths the fun way. I was so intrigued by the calculator to make me do the work more, as I got bored just doing paper and pencil work. It also helped me remember the mathematical symbols for addition, subtraction, multiplication and division. This allowed me to remember what the symbol looks like, because I was constantly looking and using the calculator.

This is quite unconvincing as a description of learning with technology, and indicates that the student's conception of the calculator was restricted to regarding it as a machine to do arithmetic. However the same student seemed to recognize a place for humans in calculation:

Calculators are made to simplify a task or solution for students. The student still have to think mentally even while using a calculator, because it is still up to the students to actually read the problem, understand what is asked, determine an appropriate mathematical equation to solve, solve the equation using the right method and then will have to interpret the answer to make sure it makes sense.

Other pre-service teachers also regarded the purpose of the calculator as restricted to calculating, and rarely connected them with credible descriptions of their learning. The following example was not unusual:

My calculator has helped me a lot because not only has it given me the answer, but it has helped me to understand how to get the answer without using a calculator because once I have the answer I can do the equation in my head until I get that answer because I know what I am looking for. Another way a calculator has helped me in my learning for this unit this semester is when we have to work out large sums preferably in our heads we can use our calculator to double check the answer.

Examples of this kind make it clear that there seems to be more to be learned here by the teacher than by the students, and there continues to be a need to help students learn about mathematical ideas by using a calculator to represent and manipulate them. In these cases, it may be that student's reflections do not indicate that they have thought deeply about things, but rather that they have completed a required assessment task, using a readily-available example of technology, but not really understood the purposes of doing so.

4.11 *Understanding mathematics?*

Many student reflections gave the impression that students recognized the importance of understanding mathematical concepts and their relationships with each other, of key importance to future (and existing) teachers. For example, when reflecting on learning about areas and volumes of rectilinear shapes, one pre-service teacher wrote:

> *I think that students need to be able to recognize the relationships between numbers and work out the formulas for themselves as opposed to just being given a formula to memorise.*

Similarly, the reflections of the following pre-service teacher make it clear that they are beginning to recognize the key role of the teacher in shaping student opinions and attitudes about mathematics. The importance of understanding in mathematics is becoming clear to this pre-service teacher:

> *I have been wondering why I am experiencing mathematical enthusiasm now as opposed to suffering it in school previously and I have come to two conclusions. The first is that teaching methodologies have changed for the better. Secondly, and even more importantly, I am motivated to understand mathematics now and I understand that I have to not only teach mathematics and answer mathematical questions of 'my future kids' but I feel I need to make sure that they, unlike me, do not grow up with an irrational and misplaced fear of mathematical problems. I can see the fun in it now and I look forward to passing it on to my students!*

Indeed, reflecting on the place of reflection itself, one pre-service teacher wrote:

I used to think that if a problem is hard to solve I would just give up on it and fall even further behind, now I have a different mindset and I have learnt to enjoy trying to figure out different ways of working things out to find it easier for myself. After writing this reflection I have been able to think more about place value and by doing this I now think that I have a better understanding of how it works and why it is necessary.

Reflections of these kinds suggest that some of the pre-service primary teachers have used the reflection process to think about their own learning, as well as about their eventual role as a mathematics teacher to support the learning of others.

5 Conclusion: Reflecting on Reflections

The various themes extracted from the reflections of pre-service primary teachers offer a window into the opportunities that might be created for both students and teachers by the use of reflective work of these kinds. It seems that the requirement to engage in reflective thinking has encouraged these students to address some issues associated with learning mathematics, some issues about the nature of mathematics and some personal issues. Time spent in such work would seem to be valuable for the learners, engaging them in the act of thinking about their own thinking.

In this case, some caution is needed about interpreting these experiences when considering teaching children in schools. The pre-service teachers described here were able (and were encouraged) to draw upon longer personal histories of learning mathematics than would usually be the case for children in schools. In addition, their mathematics unit was described by many of them as different in character from their previous studies of mathematics. Together with their relative maturity (as adults), and career-orientations, these factors may have helped the pre-service teachers to use reflections in a positive way. Similar tasks would not necessarily be the optimum way for teachers in schools to engage their pupils in reflections.

It is clear that reading the pre-service teacher reflections offers important opportunities for their teacher as well, providing a means of reflecting on the success or otherwise of activities in particular and the unit in general. Such opportunities were of course constrained by the circumstances to some extent in this case: some students regard assignment work — even work that involves personal reflection — as mostly concerned with writing what they think the teacher would like them to write. This seems to be an inevitable consequence of engaging in reflection as part of a formal assessment process.

Despite these limitations, the reflection expectations seemed to offer some valuable insights for both the pre-service teachers and their teacher (the author). In addition, in this case (although not generally for teachers), the pre-service students engaged in reflection were themselves beginning a process of learning to be teachers, and there are lessons to be learned by them in the benefits of engaging in reflective thinking, that will hopefully inform their work in the future as classroom teachers.

Acknowledgements

The author thanks all the anonymous students whose work has informed his understanding of the powers of reflection and provided much of the detailed content for this chapter. He is also grateful for constructive feedback on this paper from both Lorraine Jacob and Anne Watson.

References

Australian Curriculum, Assessment and Reporting Authority. (2012). *Australian curriculum: Mathematics*. Retrieved October 7, 2012, from http://www.australiancurriculum.edu.au/Mathematics

Bogomolny, A. (2012) *Pythagorean theorem by rearrangement*. Retrieved October 7, 2012, from http://www.cut-the-knot.org/Curriculum/Geometry/ArrangePyth.shtml

Cambridge University Press. (2011). *HOTmaths*. Retrieved October 7, 2012, from http://www.HOTmaths.com.au/

Dewey, J. (1933). *How we think: A restatement of the relation of reflective thinking to the educative process*. Boston: D.C. Heath.

Haylock, D. (2010). *Mathematics explained for primary teachers*. (4th Ed.) London: SAGE.

Lesh, R., & Zawojewskii. (2007) Problem solving and modeling. In F. Lester Jr. (Ed.), *Second handbook of research on mathematics teaching and learning* (pp. 763-804). Charlotte, NC: Information Age.

McDonald, J., & Dominguez, L. (2009). Reflective writing. *The Science Teacher, 76*(3), 46-9.

Ministry of Education. (2006). *Mathematics syllabus - Primary*. Singapore: Author.

National Council of Teachers of Mathematics. (2012). *Illuminations*. Retrieved October 7, 2012, from http://illuminations.nctm.org/

Shodor Education Foundation. (2012). *Interactivate*: *Area explorer*. Retrieved October 7, 2012, from http://www.shodor.org/interactivate/activities/AreaExplorer/

Skemp, R. (1976). Relational and instrumental understanding. *Mathematics Teaching, 77*, 20-26.

Utah State University. (2010). *National library of virtual manipulatives*. Retrieved October 7, 2012, from http://nlvm.usu.edu/en/nav/vlibrary.html

Wheatley, G.R. (1992). The role of reflection in mathematics learning. *Educational Studies in Mathematics, 23*(5), 529-541.

Wiliam, D. (2007). Keeping learning on track. In F. Lester Jr. (Ed.), *Second handbook of research on mathematics teaching and learning* (pp. 1053-1098). Charlotte, NC: Information Age.

Wilson, J. (1998). Metacognition within mathematics: A new and practical multi-method approach. In C. Kanes, M. Goos, & E. Warren (Eds.), *Teaching mathematics in new times. Proceedings of the 21st Annual Conference of the Mathematics Education Research Group of Australasia* (pp 693-700). Griffith University Print: MERGA.

Chapter 8

Reflecting on Calculation: When Drilling Becomes Fulfilling

Anne WATSON

This chapter illustrates many forms of reflection that can be informative when working on mathematics. Reflection can reveal patterns in mathematics by focusing on what is different when other things are the same, or what is the same when other things are different. Teachers can construct situations in which these characteristics are revealed. Reflections on the effects of actions can reveal underlying relations, and further reflection can be invoked to compare different representations of the same actions. Sometimes reflection can be triggered by the sudden realisation that learners had a subconscious expectation, generalisation, or assumption, which has just been challenged by new example. Again, teachers can engineer this situation. Reflection on a sequence of work can enable learners to recognise similarities at a higher, more abstract, level than they had experienced while doing the work. This can even take place during routine calculation practice with well-designed examples. Finally, reflection on the learning process itself can give learners a sense of their own progress. But mathematics is not an empirical subject. Engaging the reflective mind can enrich understanding and transform knowledge, but deductive reasoning has to be engaged to retread the pathways of action through reasoning.

1 Introduction

The desire for students to reflect on learning is stated in many national curricula and since the 70s the phrase 'reflective learning' has gained currency internationally. Sometimes this is taken to mean that students who think back over what they have done improve their memory of it, or their sense of engagement with it, or get a better sense of their overall progress. However in this chapter I am going to develop a more cognitive use of the word based on theories that taking an overview of work just done can reveal generalisations that might otherwise be overlooked in the sweat of 'doing' the tasks. Reflection, at its most cognitively useful, can reveal patterns in mathematics by focusing on what is different when other things are the same, or what is the same when other things are different.

Teachers can construct situations in which various characteristics are revealed. For example, reflection on the effects of actions can show underlying relations between what is done and what happens as a result. Reflection can be invoked to compare different representations of the same actions, so that meaning and representation are not irrevocably connected. Sometimes reflection can be triggered by the sudden realisation that learners had a subconscious expectation, generalisation, or assumption, which is just been challenged by new example. Reflection on a sequence of work can enable learners to recognise similarities at a higher, more abstract, level than they had experienced while doing the work. This can even take place during routine calculation practice with well-designed examples. Finally, reflection on the learning process itself can give learners a sense of their own progress. Again, teachers can engineer this situation.

But mathematics is not an empirical subject. It cannot be understood from merely generalising patterns which have been generated by the work done. Indeed a common cause of error in mathematics is using methods or making assumptions that are based on over-generalisation (Avital & Barbeau, 1991) such as using techniques for transforming equations to transform inequalities, or putting a zero 'on the end' when multiplying by ten. That students' errors are so often based on applying patterns which work on some other domain is evidence of the generalising mind. Engaging this reflective mind, that seeks invariance and variation, can enrich understanding and transform knowledge, but deductive reasoning has to be engaged to retread the pathways formed by mathematical action. For mathematics to be understood, the generative and generalising pathways that result from working on methods and facts have to be retrod through reasoning.

Various authors have found ways to describe the need to retread empirical experience while engaging the reasoning mind. In geometric thinking, the van Hieles devised a developmental description which starts with visual and physical experience (1986). New objects have to be experienced and new situations analysed to learn their components, construct definitions, and know how components relate to each other. Finally, these relations become new objects in themselves about which theorems can be stated. In mathematical experience more widely, Mason talks about the aim of manipulation being to 'get a sense of' a mathematical object, and the sense being finally articulated in words, symbols or diagrams which themselves are new objects (e.g. in Floyd, Burton, James, & Mason, 1981). I also want to mention Wertheimer, a foundational thinker about gestalt, who suggested that our sense of the whole of something includes not only its components but also the ways they embody and represent fundamental relations (Wertheimer, 1961). In all of these traditions, the act of looking at the whole, and seeing through it to the integral relations, is central to learning about new ideas. In contrast to this approach, it is still the case that for many children learning mathematics is not about whole objects, but about sequences of small repetitive tasks. The act of looking back over these small repetitions can be evoked to generate learning, but often in the classrooms I observe that this 'looking back' only takes the form of checking answers.

An example I have used for over ten years with prospective teacher applicants about sixty times a year is a picture of a child's work on an exercise involving multiplying fractions. One of the questions is:

$$\frac{2}{3} \times \frac{5}{8} =$$

and the following question is:

$$\frac{1}{2} \text{ of } \left(\frac{2}{3} \times \frac{5}{8}\right) =$$

The people I use it with are adept at commenting on various aspects of the second question, such as whether a learner would understand what 'half of' means and whether they should have to do the brackets first or not. Very few notice that the second question is merely half of the first, and the few that comment on this seem to think it is a sophisticated thing to notice. This assumption that individual questions are free-standing in typical textbook exercises has caused me to think about the structure of question sets in textbooks more creatively (Watson & Mason, 2005).

2 Where to Look for Patterns

For young children a sense of the structure of a whole mathematical idea is likely to be through *physical actions*, so that abstract ideas develop from their actions on the objects they are given, whether they are physical objects or symbolic objects appearing as pen on paper. For example, dropping down one line every time you add ten on a hundred square is physical, and enacts the action of adding tens so that the units digit remains untouched. When multiplying using repeated addition, the grouping and collating of counted objects gives a generalisable sense of multiplicative patterns. Manipulatives such as Numicon ™ which embed properties of number (in Numicon's case this is even and odd numbers) can give children a physical sense of parity qualities of number.

Patterns, particularly patterns of the *effects of similar actions*, provide evidence of relations, which is why pattern-spotting is important in mathematics. Even simple one-dimensional patterns give us such evidence — we seek to describe how they develop and show how the development can be continued. Patterns of *action and effect*, which help us predict behaviour, combine action and pattern seeking to understand operations.

All of these use *inductive reasoning*, that is reasoning which assumes patterns will continue. To engage in these sorts of reasoning, learners have to reflect on what they have produced, to look for patterns, and also to reflect on the effects of their actions, so they can make predictions, check their work, invent short cuts and so on. A simple example of this is the following exercise, based on what we have found in several Finnish textbooks for young children:

$17 - 9 =$

$27 - 9 =$

$37 - 9 =$

$47 - 9 =$

...

Children might start by using manipulatives, number facts, or counting strategies to find the first and second answers, but after that reflection on the action of subtracting 9, and the effect of changing the units digit to 8, can lead to a local generalisation which enables them to complete the exercise by using number structure rather than calculation methods.

Patterns are not limited to visual patterns, but also to patterns of *symbolic meaning*, such as sequences of square numbers, multiples of five, and so on. Furthermore, patterns of *similarity and difference across nearly similar objects*, or objects that differ only in a few aspects, also count as evidence of mathematical relationships. For example, the observation that the diagonals of members of a collection of rectangles all bisect to each other can lead to an inductive generalisation that this is a property of all rectangles. To be fully sure requires deductive reasoning, and a learner who has already made this statement for themselves and based on their experience is in a better position to (a) remember this property and (b) understand the need for the logical proof.

3 Shifting to Deductive Reasoning

However useful patterns are, in mathematics we also need another form of reasoning as well, reasoning that starts with definitions, properties and implied relations, in which logical implications and mathematical coherence take priority over observed patterns. This reasoning starts from describing the fundamental characteristics and relations of a class of objects, and then constructs statements about inclusion in such classes, and inferences about further properties. I am going to illustrate this in the context of learning about the multiplicative relationship, because it is fundamental to understanding and using ratio, proportion, percentages.

Before I present tasks, however, it is important to observe that any task for which an easily-obtained answer is the desired endpoint is unlikely to lead to reflection on its own except for learners to reflect on their ability to obtain answers. Yet reflection is necessary to observe structural patterns, because one case, one answer, one worked example, does not provide the data for seeking pattern and, hence, relations. In mathematics, reflection requires a further dimension; to reflect a two-dimensional shape we have to move it through the third dimension. A similar thing can be said about the psychology of reflection on action and on the effects of action. Left to ourselves we may not have the next dimension to use to help us reflect on less obvious features. For example, learners might reflect on a task in terms of what they wrote down and whether they got right answers or not, but a teacher can intervene to take reflection to a new level of abstraction, such as by asking them to say what was the same or what was different about two of the questions. This intervention can draw attention to features which learners may not have thought of seeking out for themselves. It takes well-designed tasks, good teaching and deliberate talk to promote reflection that brings new, higher level perspectives. This is what Vygotsky was talking about when he spoke of a zone of proximal development (1978), and also what Bruner meant when he suggested that learning new ways of thinking could be achieved through scaffolding (e.g. in Wood, Bruner, & Ross, 1976).

It is frequent, in mathematics education publications, to claim that mathematical thinking and reflection can be most easily triggered in non-routine tasks, and tasks that promote rich, diverse, mathematical activity, such as solving unfamiliar problems. In the following sequence of tasks, I am trying to show that repetitive, practice-type, tasks and sequences of tasks can provide a structured context for reflective engagement leading to deep learning about the multiplicative relation. In fact, I could go so far to say that sometimes such tasks are the best way to do this. They can evoke insights which are achieved through a kind of fluency which gives learners a sense of extendable and universal method. It is even the case that repeated written actions and repeated layouts can give a physical sense of pattern and structure.

These tasks all revolve around the multiplicative relation, by which I mean the connection between a, b and c expressed as $a=bc$ or $\frac{a}{b}=c$, or $\frac{a}{c}=b$. Full understanding of this goes beyond knowing multiplication tables, and beyond the idea of repeated addition or arrays (which are a spatial representation of repeated addition). Deep understanding of the relationship is necessary for the fluent use of ratio and proportion in mathematics, in other subjects, and in everyday contexts.

3.1 *Task 1*

The following exercise is practice in reversing multiplication facts to find quotients, i.e. using knowledge of tables to do division:

$$60 \div 1 = 60$$
$$60 \div 2 = 30$$
$$60 \div 3 = 20$$
$$60 \div 4 = 15$$
$$60 \div 5 = 12$$
$$60 \div 6 = 10$$
$$60 \div 7 = ??$$

The sudden appearance of one that 'doesn't work' provides the trigger for discussion. Maybe the word 'factor' could be introduced for those that 'do work'. What could the word 'divide' mean in this context? It does not seem to be about sharing out by counting, nor about the inverse of a known multiplication fact. Instead it provides the jumping-off point for deeper understanding because the task has offered $60 \div 7$ as a contrast to the unspoken expectations from other cases. The power of the task to evince surprise depends on learners having expectations which have been created in the early stages of the task. Nothing will happen at all if the teacher emphasises right answers as the end point, rather than right answers giving data for reflection; nothing will happen if students do not know multiplication facts and therefore worked out each separate calculation using the same method that will then, with extension, 'work' for dividing by seven.

There is more: suppose the teacher then asks students to compare: '$60 \div 3 = 20$' and '$60 \div 6 = 10$'. What could be noticed? Maybe she introduces the word 'double' to draw attention to the three and the six. What can then be said? Suppose learners conjecture that doubling the divisor halves the quotient — there are other pairs of cases that can extend this idea. Reflecting on similarities and differences in answers leads to the possibility of a deeper understanding of whole number multiplication, as well as new meanings for division.

3.2 *Task 2*

Think of a number Multiply it by 10 Add 6 Divide by 2 Add 2 Divide by one more than your original number The answer is ……

Here, the pattern to be found is that different inputs put through the same processes yield the same answer. This is not usually the case in life or in mathematics, so reflection on the work of the whole class can trigger intrigue. Central to understanding this process is the inverse relation between multiplication and division, although this is hidden by adding in some distracting numbers. What happens to multiples when we add things before or after dividing can be explored in simpler cases, such as 'multiply by a number of your choice; then divide by the same number'.

Then 'multiply by 10; divide by 2; divide by 5'. Nothing will happen at all unless the teacher invites students to compare their starting points.

3.3 *Task 3*

$$
\begin{aligned}
1 \div 9 &= 0.111 \ldots \\
2 \div 9 &= 0.222 \ldots \\
3 \div 9 &= 0.333 \ldots \\
4 \div 9 &= 0.444 \ldots \\
5 \div 9 &= 0.555 \ldots \\
6 \div 9 &= 0.666 \ldots \\
7 \div 9 &= 0.777 \ldots \\
8 \div 9 &= 0.888 \ldots \\
9 \div 9 &= 0.999 \ldots
\end{aligned}
$$

In a written text it is impossible to re-create the sense of impending interest that can happen as the questions are tackled one by one. Learners will soon abandon their own calculations and rely on pattern continuation, with or without justification, to generate answers. Discussion of whether it is right to do so, and why, for example, $3 \div 9$ gives three times the answer for $1 \div 9$, can follow. Learners can ask their own questions, make their own observations and conjectures, by reflecting on the work done and picking out possible patterns and relations. Of course, the lurking excitement is the counter-intuitive result that $9 \div 9 = 0.999 \ldots$ when usually it is written as 1. Indeed, when this sequence was revealed in a presentation to primary teachers there was audible excitement from the audience, not all of whom were familiar with the counter-intuitive final item. The power of division to present this fact is intriguing for learners, but is more about the potential for the multiplicative relation to promote new mathematical ideas once it is understood then deepening understanding of the relation itself.

3.4 *Task 4*

In this task, learners are asked to decide in advance whether some given numbers are divisible by thirteen and, using long division if necessary, verify their expectations.

Guessing, learners usually decide that a 'nice neat' number like 222222 cannot be divided by a messy number like 13. Those who decide otherwise usually do so because they have a feeling that they might be caught in a trick. The answer is 17094. Given that knowledge, they are then asked about 444444 and 666666 as potential multiples of thirteen. These questions coming next draw attention to a property of division, that multiplying the dividend by k has the effect of doubling the quotient by k. There is no need to divide by 13 because adapting the first answer is easier and moreover is an expression of the Fundamental Theorem of Arithmetic.

$$222222 \div 13 = 17094$$
$$444444 \div 13 = 34188$$
$$666666 \div 13 = 51282$$

Next, learners are asked about 777777. This often produces the expectation that it is *not* divisible by 13 because of the inductive generalisation that strings of even numbers *are* divisible. Quick guessing often excludes odd number strings, yet they *are* divisible and learners have been deliberately misled by the first three questions. Returning to the division algorithm shows this, so the 'fulfilling' part of doing the calculation in this example is that an expectation is challenged. There is also something ready to be deduced about the divisibility of 111111 by thirteen.

$$777777 \div 13 = 59829$$

Readers can decide for themselves the expectations and intrigues that are generated by the final three questions below. Moreover, such a sequence of dividing tasks can trigger interest in dividing by other 'nasty' numbers and learners can try to make up similar misleading examples. Doing the division themselves, and understanding relations between the three numbers involved, gives clues about how these can be constructed.

101010101010 ÷ 13 = 7770007770
131313131313 ÷ 13 = ?
121212121212 ÷ 13 = ?

I observed the following procedural task being done in school by a group of rather disaffected 13 year olds triggered my interest in what can be achieved with standard drilling tasks.

3.5 *Task 5*

Learners are asked to complete as many lines of a grid as they can. The bold entries are what they are given and the non-bold entries indicate how the class started filling it in.

1 x 7 = 7	**7 x 1 = 7**	**7 ÷ 1 = 7**	**7 ÷ 7 = 1**
2 x 7 = 14	**7 x 2 = 14**	**14 ÷ 2 = 7**	**14 ÷ 7 = 2**
3 x 7 = 21	7 x 3 = 21	21 ÷ 3 = 7	21 ÷ 7 = 3
4 x 7 = 28	7 x 4 = 28	28 ÷ 4 = 7	28 ÷ 7 = 4
5 x 7 = 35	7 x 5 = 35	35 ÷ 5 = 7	35 ÷ 7 = 5
6 x 7 = 42	7 x 6 = 42	42 ÷ 6 = 7	42 ÷ 7 = 6
7 x 7 = 49			
8 x 7 = 56	7 x 8 = 56	etc.	
9 x 7 = 63	etc		
10 x 7 = 7			

Not only was there little that was fulfilling about completing the table, but the methods they used failed to engage with the meaning of multiplication as the task designer had intended. Nearly all of them filled out all the 7s vertically, then all the natural numbers vertically, and finally discussed products of seven among themselves to complete the remaining gaps. The line that began 7x7 caused a little confusion, possibly because it is obvious what to write if learners are thinking about the meaning, but if they are only seeing it as a gap filling exercise it is not so obvious.

To make the task more fulfilling by initiating a reflective move I asked some of them to predict the contents of the row that would start 7x35, but not to work out the product. All of them could do this by following a pattern of entries, using the other rows as templates. This shift of attention is not hard to make, but had not occurred to them. As I indicated earlier, reflection requires a knowledgeable person to scaffold a new perspective. I call this new perspective 'going across the grain' where the original work was carried out 'with the grain' (Watson, 2000). As with sawing wood, going with the grain is easy and splits the wood; going across the grain is hard and reveals structure — in this case the structure of the multiplicative relationship expressed through commutativity and inverse relationships. The role of the teacher here was to direct attention away from the order in which the task had been completed towards observing and expressing the underlying relation, and maybe generalising it in symbols.

3.6 *Task 6*

Think of a number
Multiply it by 12 and divide by 3
Now multiply the output by 3 and divide by 12

Think of a number
Multiply it by 15 and divide by 5
Now multiply the output by 5 and divide by 15

Think of a number
Multiply it by a and divide by b
Now multiply the output by b and divide by a

In Task 6 both the layout and the words, if read out loud, indicate repetitions of actions by that have a similar structure. The final version aims to generalise that structure by de-emphasising the numbers and emphasising the actions. The task design shifts of attention from doing the arithmetic to symbolising the process, and therefore takes a step towards explaining the effects of the actions.

4 Reflecting on the Design of the Tasks

All the tasks I have presented have features in common. Each of them starts with arithmetical operations to be performed. That is the 'drilling' in the title of this paper. But each task contains features which shift the learners' attention towards a general property of number relations. In each task there is something to be noticed and discussed in lessons. Many of these noticeable features can be spotted by learners as they do the tasks. Others require a teacher to direct learners' attention across the grain. Tasks which start with drilling can end up being a mathematically fulfilling in this way, and have the added advantage that learners already have an investment in making sense of the outcomes of their own work.

Task 1 introduced a sudden disruption to a smooth sequence of work, and hence the need for a new understanding of division. Task 2 offered a classroom full of different examples which all, possibly surprisingly, gave the same answer; hence creating a need for explanation. Task 3 ostensibly offers practice in a non-linear exact division, but the answers so obviously form a pattern that there is plenty to discuss about ninths. Task 4 focuses further on division by presenting some related results that require some algorithmic effort; the effort is rewarded with a deeper understanding of the effects of changing the dividend and the divisor multiplicatively. Task 5 tempts learners to fill gaps using patterns, and requires the teacher to direct attention towards the whole multiplicative relation. Task 6 steps towards an algebraic explanation-proof of the effect of inverse operations.

These tasks achieve these aims by the careful control of variables within the repeated examples. They offer in patterns in physical actions, patterns in the effects of the actions, patterns in the relationship between action and effect, patterns in the symbolic relationships, and patterns of similarity and difference. They therefore invite inductive reasoning towards particular generalisations which can move learners towards a more abstract understanding, in this case of the multiplicative relationship.

Underlying all this is the belief that learners can and will:

- Reflect on what they have produced, to look for patterns
- Reflect on the effects of their actions, so they can make predictions, check their work, invent short cuts
- Will learn that reasoning from experience can lead to false conjectures so they need logical reasoning too.

Teachers therefore have a responsibility to:

- Provide sequences of tasks that reveal mathematical patterns through of the controlling variables and variability
- Encourage learners to reflect on the effects of their actions, so they can make predictions and develop conjectures
- Provide tasks that show how reasoning from experience can sometimes be misleading.

5 Summary

In this chapter I have shown that well structured tasks and pedagogic prompts can generate a special kind of mathematical reflection which involves noticing patterns, similarities and differences in repetitive work work that has just been carried out. The conceptual understanding arises from the actions and effects of work which traditionally might only be seen as drilling. However, reflecting across the work can transform knowledge about calculation and show how methods and concepts interact. Engaging the reflective mind on the apparently menial tasks can enrich understanding.

There is a further bonus which is that learners themselves can identify the pathways which led from drilling to the fulfilment that comes about from understanding mathematical ideas.

References

Avital, S., & Barbeau, E. (1991). Intuitively misconceived solutions to problems. *For the Learning of Mathematics, 11*(3), 2-8.

Floyd, A., Burton, L., James, N., & Mason, J. (1981). *EM235: Developing mathematical thinking*. Milton Keynes: Open University.

van Hiele, P.M. (1986). *Structure and insight: A theory of mathematics education*. New York: Academic Press.

Vygotsky, L. (1978). *Mind and society: The development of higher psychological processes*. Cambridge, Mass.: Harvard University Press.

Watson, A. (2000) Going across the grain: Mathematical generalisations in a group of low attainers. *Nordic Studies in Mathematics Education, 8*(1), 7-20.

Watson, A., & Mason, J. (2005) Seeing an exercise as a single mathematical object: Using variation to structure sense-making. *Mathematical Thinking and Learning, 8*(2), 91-111.

Wertheimer, M. (1961). *Productive thinking*. London: Tavistock Publications.

Wood, D.J., Bruner, J.S., & Ross, G. (1976). The role of tutoring in problem solving. *Journal of Child Psychology and Psychiatry, 17*(2), 89-100.

Chapter 9

Developing Reflective Learners Through Solving Non-Routine Problems

Marian KEMP

Students who are taught by the method of demonstration with explanation followed by practice exercises may believe that the mathematics they learn in the classroom is just a set of arbitrary rules that they learn from the teacher. In this context students may be actively engaged in their learning but it is likely to be inadequate for deep, relational learning to occur. The literature indicates that students' understanding of mathematical concepts can be improved through a problem solving approach that incorporates students reflecting on the mathematical activity that is being undertaken. Students can help to develop their reflective thinking through the asking of cognitive questions of themselves while problem solving, preferably in a community of learners accustomed to critically questioning assumptions and methods. This chapter suggests that opportunities can be provided in the classroom to develop and nurture reflective learners through the choice of suitable open-ended or non-routine problems for students to engage with. Some criteria for choosing problems are provided, together with some examples of problems meeting these criteria.

1 Introduction

Traditionally many teachers have believed that students' learning takes place as a result of clear explanation and practice exercises, based on their own experiences at school. Indeed, according to Artzt, Armour-Thomas, and Curcio (2011):

> One widely held belief about teaching is that the role of the teacher is to transmit mathematical content, demonstrate procedures for solving problems, and explain the process of solving sample problems. Although there may be some instances where such an approach may be appropriate, there is much more to teaching for student understanding (p. 5).

In line with this view that there is more to teaching mathematics than a solely didactical approach, this chapter is concerned with providing suitable classroom activities to engage students in problem solving and developing as reflective learners, who can be critical thinkers and reflect on the reasonableness of their solutions. Schoenfeld (1987, p. 197) was concerned that "many students come to believe that school mathematics consists of mastering formal procedures that are completely divorced from real life, from discovery and from problem solving". He cites the example from the Third National Assessment of Academic Progress examination where students are asked to find the number of buses required to transport 1128 soldiers by bus if each bus can take 36 people. Of the 45 000 secondary students who completed the question 70% completed the division calculation correctly. Of those students 29% gave the answer that the number of buses needed is 31, remainder 12; 18% said the number of buses needed is 31, 23% said the number of buses is 32, the correct answer (p. 196). Unfortunately an emphasis on rules and memorisation, divorced from reality, is still the case in some classrooms.

In contrast, acknowledging that "mathematical problem solving is a critical component of a Mathematics Education and should not only be a goal in mathematics learning but also a way to acquire new content" (NTCM, 2000), the Singapore Mathematics Syllabus (Ministry of Education, 2006) puts problem solving at the heart of the curriculum. It states that:

> Mathematical problem solving is central to mathematics learning. It involves the acquisition and application of mathematical concepts and skills in a wide range of situations, including non-routine, open-ended and real-world problems. The development of mathematical problem solving ability is dependent on five inter-related components, namely, *Concepts*, *Skills*, *Processes*, *Attitudes* and *Metacognition* (p. 2).

This chapter will give an overview on problem solving and then consider the role of solving non-routine and open-ended problems in the development of reflective learners and the integration of metacognitive questioning in the process.

2 Overview of Problem Solving

According to Martinez (2006), "Problem solving can be defined simply as the pursuit of a goal when the path to the goal is uncertain. In other words, it's what you do when you don't know what you are doing" (p. 697). Indeed, problem solving is not restricted to the mathematics classroom; it can be a daily activity in a variety of situations where decision-making is required, like purchasing a car, building a house or planning a holiday, many of which use some mathematical knowledge and skills. Mathematical problem solving involves thinking mathematically in a broader sense although content and skills are a necessary component. Ideally, problem solving helps students to think creatively or critically; develop their problem solving ability; to learn techniques in modelling and develop analytical reasoning skills (Schoenfeld, 1992).

Over the years "problems and problem solving have had multiple and often contradictory meanings" (Schoenfeld, 1992, p. 342). Some people see a 'problem' as a task to be done while others see a problem as a difficult or perplexing question. The former interpretation is a somewhat traditional one, where the teacher demonstrates the mathematical technique or concept using a problem as a vehicle, and then the students complete routine exercises (sometimes called word problems) as practice towards mastering the technique or understanding the concept. The latter interpretation acknowledges that a problem has no obvious solution to the student. In this mode a problem is a task for which: (i) the individual or group confronting it wants or needs to find a solution, (ii) there is not a readily accessible procedure that guarantees or completely determines the solution and (iii) the individual or group must make an attempt to find a solution (Lester, 1983, pp. 231-232). However, "... being a problem is not a property of a mathematical task. Rather it is a particular relationship between the individual and the task that makes the task a problem for that person. A problem for one person may simply be an exercise for another person" (Schoenfeld, 1985, p. 74). So their teacher, who has a good sense of the students' mathematical knowledge and skills, should carefully select the problems for students to solve so they are not too hard for them so as to be unsolvable and not too trivial so they are just exercises.

In the context of problem solving Stanic and Kilpatrick (1989, p. 1) point out that, "problems have occupied a central place in the school mathematics curriculum since antiquity but problem solving has not." However, since George Polya wrote *How to Solve It* in 1945 there has been an interest in the processes that problem solving involves and attempts to introduce it into the school curriculum have been more or less successful depending on the local mathematics curriculum. His description and elaboration of four steps for problem solving (i) understanding the problem (ii) devising a plan (iii) carrying out the plan and (iv) looking back. (Polya, 1990, xxvi-xxvii) indicate that reflective thinking is important throughout the process.

Since the work of Polya, other educators and researchers have described problem solving with varying, but similar, definitions. Schoenfeld (2011, pp. 3-4) describes a framework of problem solving in terms of four components: *knowledge base* — what mathematics do they know; *problem solving strategies* — tools or techniques to make progress on problems they don't know how to solve; *monitoring and self regulation* — managing resources including time, at their disposal; *beliefs* — an individuals sense of mathematics, the context, themselves and what they choose to do.

In addition, Mayer (1998) suggests that in broad terms problem-solving can be thought of in terms of three main kinds of skills: cognitive skills (concepts and skills in domain-specific knowledge of mathematics); metacognitive skills (metacognition and processes — the ability to control and monitor cognitive processes) and motivational skills (attitudes, beliefs, feelings and interest in the problem). In this categorization Mayer refers to these three as: skill, metaskill, and will. He points out that in the solving of non-routine problems, students need metaskill because it "manages and coordinates the other components" (p. 51).

Out of these components the least generally understood term is metacognition, first coined by Flavell in 1976 and his definition is now generally accepted:

> Metacognition refers to one's own knowledge concerning one's own cognitive processes and product or anything related to them, e.g., the learning-relevant properties of information or data ... Metacognition refers, among other things to the active monitoring and consequent regulation and orchestration of these processes in relation to the cognitive objects on which they bear, usually in the service of some concrete goal or objective (Flavell, 1976, p. 232).

Since then there has been considerable interest and work in this area and whilst there is some consensus in the literature that metacognition, is 'thinking about thinking', and that it is important, there are different perspectives on what it comprises and how it can be developed (Wiliam, 2007, p. 1079). According to Flavell (1979, p. 906), the term metacognition is an umbrella term that includes "knowing what one knows ... what one can do ... and what one knows about one's own cognitive abilities".

Even though content knowledge is important in a problem-solving context, it is equally important that students are able to manage their skills and their thinking. So, "what is the role of metacognitive behavior in problem solving?" was a question asked by many mathematics educators from the early 1980s (Schneider & Artelt, 2010). Over the years it became apparent that mathematics teachers were finding it difficult to design lessons to help students solve problems with more than one step (Kramarski, 2008). It was thought that the concept of metacognition would be a useful one to improve this situation (Lester, 1983). Brophy (1986) endorsed this view, suggesting that teachers should talk about which skills are relevant, how to use the skills and why and when to use them.

There have been a number of intervention programs designed to investigate relationships between the development of metacognition and success in problem solving, for both low and high achievers. These have generally shown that metacognitive instruction leads to better problem solving. (e.g., Hembree, 1992; Mevarech, 1999; Mevarech & Kramarski, 1997; Schoenfeld, 1992).

One such program for secondary school students developed in Israel is called IMPROVE which stands for: **I**ntroducing new material, **M**etacognitive questioning, **P**racticing, **R**eviewing, **O**btaining mastery on higher and lower cognitive processes, **V**erification, and **E**nrichment and remedial. IMPROVE used metacognitive questioning: *comprehension* questions (What is the problem all about?), *connecting* questions (How does the problem relate to others already solved in the past?) *strategic* questions (What kinds of strategies are appropriate for solving the problem and why?) and *reflection* questions (Does the solution make sense? Can the problem be solved in a different way?) as part of the metacognitive instruction (Mevarech & Kramarski, 1997; Mevarech & Amrany, 2008).

Evaluations of the IMPROVE program showed that the program had the potential to improve students' problem solving abilities and that cooperative learning on its own was not sufficient to ensure problem-solving success. The students performed better when they worked in groups and incorporated metacognitive questioning to help them during and after the problem solving activity. The researchers found that the program had positive effects on students' planning, comprehension and reflective skills (Schneider & Artlet, 2010).

A second program, also from Israel, examined the specific area of developing reflective thinking. Hershkowitz and Schwarz (1999) report on the role and development of reflection in a rich mathematics-learning environment. This *Function Project* is a yearlong course for Grade 9 students learning geometry and is based around problem situations. The students have access to graphics calculators, they worked in small groups and interacted with their peers and their teacher.

One of the problems involved a 'real life' problem concerning wooden boxes for the overseas transport of goods in the most efficient and economical way, minimizing the amount of wood that would be needed for a particular volume given a maximum height because wood is expensive. When they have worked on the problem the:

> Students write group or individual reports, in which they are encouraged to reconstruct, compare, and critique the steps they and others went through while solving the problem. They are asked to report dialectic processes between different hypotheses: how and why a hypothesis was made or discarded, why another one was raised, and so forth (p. 68).

This led to opportunities to critique each other's work so the students became a community of learners used to reporting and criticizing.

As part of the program the teacher played a major part in "supporting reflective discourse in the room" through interaction with the students, asking questions and where relevant suggesting extra tasks. Intervention programs of this kind make it clear that reflection is an important aspect of problem solving and they indicate that the development of reflective learners can be achieved through suitable meatacognitive questioning as part of the problem solving process.

3 Developing Reflective Learners

In order to be able to nurture reflective learners teachers need to understand that reflection is not just simply looking back. In reviewing Dewey's work, Rodgers (2002) states that, "reflection is a particular way of thinking and cannot be equated with mere haphazard 'mulling' over something" (p. 849). In Mewborn's (1999) research on reflection in pre-service teacher education she noted that whilst the term reflection is used in a variety of ways there are three features of reflection that are in common with them all. Firstly, reflection is different from recollection or rationalization; secondly that action is an integral part of the reflective process and thirdly that reflection is both an individual and shared experience (p. 317). The importance of reflecting on 'something' is emphasized by Wheatley (1992), who argues out that "reflection plays a critically important role in mathematics learning and that just completing tasks is insufficient" (p. 529). So, how can we approach the development of reflection in the mathematics classroom?

In the process of reflection the mathematical activity that the students are undertaking becomes the focus of discussion. When solvers reflect on solution activity, "they make their activity an object that can be examined" (Wheatley, 1992, p. 536). This necessarily means that students need to slow down and think about what they are doing. As Shimizu (2009) has pointed out, this is a natural part of Japanese Lesson Study as the students engage in a large proportion of the lesson reflecting on their own processes and those of their fellow students.

Research suggests that there are particular kinds of questions that can be asked that encourage students to engage in reflective thinking and learning. The teacher can help students to become familiar with the kinds of questions that will help them with their problem solving: Schoenfeld (1992, p. 397) suggests that metacognitive questions such as *What (exactly) are you doing? (Can you describe it precisely?), Why are you doing it? (How does it fit into the solution?), How does it help you? (What will you do with the outcome when you obtain it?)* are questions to ask students on a regular basis.

Hershkowitz and Schwartz (1999) indicate that IMPROVE uses questions such as: *Does the solution make sense? Can the problem be solved in a different way?* Similarly, Polya (1990) suggests: *Can you derive the result differently? Can you see it at a glance? What would you gain by doing so?* Students can become fluent at using these kinds of questions, automatically answering them for themselves without teacher prompts if sufficient attention is paid to explicitly using them in the classroom. As Schoenfeld (1992) reported, "students begin to defend themselves against them by discussing the answers to them in advance. By the end of term this behaviour had become habitual". (p. 397)

Polya (1990) suggests that teachers must be careful in their kinds of questioning not to make them too specific because there are inherent dangers in doing so. For example he describes a problem where the student needs to *'Find the diagonal of a rectangular parallelepiped of which the length, the width and the height are known'*. While students are working he suggests that the question *'Do you know a related problem?'* is appropriate whereas *'Could you apply the theorem of Pythagoras'* is not. He argues that if the student is nearly at the solution that might help but if he is not close to the solution it may not be much help. Indeed, this could give the secret away and leaves little for the student to do. Even if the student understands the question he may have no idea at that stage of how the teacher came to the question and it appears as a "rabbit out of a hat" (p. 22); consequently, such a question seems unlikely to help students become more reflective.

So, when teachers are aiming to develop students as reflective learners while engaged in problem solving they need to consider the kinds of questions they will ask, the arrangement of the students in pairs or groups and the content of the task itself based on their knowledge of the students.

4 Choosing Problems to Develop Reflective Thinking

There has been much discussion and agreement in the literature around authentic and rich tasks as being suitable vehicles for learning mathematics. (e.g., Masingilia, Davidenko, & Prus-Wisnioska, 1996; Masingilia & de Silva, 2001; Zevenbergen, 1997). This is also the case with the problem solving research for which researchers using authentic tasks have found them to be appropriate for developing students' reflective thinking (e.g., Kramarski, Mavarech, & Arami, 2002; Hershkowitz & Shwarz, 1999).

It has been proposed by these researchers that context affords motivation and interest in the topic and that it demonstrates the links between school mathematics and the 'real world' which enable students to transfer their skills and knowledge to other situations. However, care must be taken in thinking about the problems as the context can "act as distracters or even barriers to understanding" (Boaler, 1993, p. 14). Her example concerns the situation where students were learning about fractions and the contexts were 'loaf' and 'cake'. A student viewed the concept of the cake as being divisible into six or eight pieces but to the student a loaf came already sliced, which added to the complexity of the problem as it meant dividing the bread slices into parts of slices.

Boaler (1993) also points out that in most real world examples that are set in schools the "students are expected to engage in a task as though it were real ignoring factors that would be pertinent in the "real life" version of the task" (p. 14). However, students do learn to work within the constraints that they are given, even if they are somewhat unrealistic. Boaler (1993) suggests that a good way to start is with a context that is open enough for the students to follow their own directions. This is worth striving for even although sometimes it is hard to see where this fits into the time available. While providing open-ended problems to challenge and motivate the students it is important to help to reflect not only on the method and accuracy of the solution but also on the mathematics underpinning the various possible solutions and approaches.

Many educators (e.g., Clement & Ellerton, 1991; Clarke, Waywood, & Stephens, 1993; Hershkowitz & Schwarz, 1999; McDonald & Dominguez, 2009; Moon, 2004;) recommend that students write in a journal about their learning. Clarke *et al.* (1993) report that, when asked about the value of their writing, students gave answers such as "because it helps me ...; to help me to learn...; to be able to explain what I think..." (p. 241). However, they point out that the development of reflection needs to be scaffolded so that the journal entries are not just descriptive.

4.1 *Guidelines for selecting problems*

Based on the findings in literature, on the use of authentic and non-routine problems in the classroom, the author suggests that the following six criteria might serve as a useful guide to the kinds of characteristics of problems that might be chosen to help students develop their reflective thinking. The problems should:

(i) be unfamiliar or non-routine to the students;
(ii) have different possible approaches or methods of solution;
(iii) require some thinking about constraints or assumptions;
(iv) be appropriate to the level of mathematical knowledge and experience of the students;
(v) be potentially interesting and motivating to the students; and
(vi) require justification or explanation of the reasons for the chosen approach and solution.

While it might seem difficult to find problems that meet all of these criteria simultaneously, each of them seems to be necessary. The choice of context and mathematical content for problems can be related to the topics the students are studying or prompted external factors such as a school fair or a class party.

The range of possible problem types can be placed on a continuum of closed at one end to open-ended problems at the other, with levels of scaffolding from structured hints to no hints at all. The kind of problem chosen along this continuum will depend on the problem solving experience and sophistication of the students. Singapore teachers have a range of good resources available to them to help develop reflective learners in the context of problem solving. Toh, Quek, Leong, Dindyal, and Tay (2011) have produced *Making mathematics practical: An approach to problem solving,* which has examples of a range of problems with suggested ways of tackling them in the classroom and helpful steps and hints for teachers.

Kaur and Yeap (2009) in their publication *Pathways to reasoning and communication in the secondary school mathematics classroom* provide a range of problem solving situations, with varying levels of scaffolding. For example, Chapter 3, *"What would you do?"* has a range of problems for which students choose from a range of cellular phones, holidays and movie tickets. Each problem provides the students with a range of three costings for alternative choices. One example is *Calling plans*: *Your parents have agreed to get you a cellular telephone for your birthday. Compare the following call plans, and choose the one that is best for you. Note that all the plans have the same cellular telephone.*

The plans are given with costs of $109, $73.50 and $40.50 per month and include different components. The students are asked (i) which calling plan makes the most sense for you? and (ii) explain your choice.

Teachers are provided with a teaching plan:

- Present the problem to the students.
- Have students read the problem individually.
- Engage the whole class in discussion and check for the comprehension of the problem.
- Create an awareness of the need to 'make assumptions' as the problem is somewhat open-ended.
- Tell the students that they must defend or justify the solution they choose.
- Ask a few students to present their solutions.
- Engage the whole class in examining the solutions presented.
- Emphasize the non-uniqueness of the solution to the problem.

This process can go a long way to helping teachers to nurture reflective learners. While using a more scaffolded approach such as this, students can learn some of the aspects of developing reflective learning in problem solving in a secure space. The scaffolding itself should help them to learn the kinds of approaches that can be made. Ideally, the next step would be to reduce the scaffolding, moving towards more open-ended problems for which students have to understand the question for themselves, decide on their approach, make clear their assumptions, make a plan, seek out the relevant information and carry out their plan.

In the case of the *Calling plans* example, one way to reduce the scaffolding might be to simply omit the information provided regarding the cellular phone plans. External information of this kind is most likely to be needed when the problems are authentic, real-life problems so that current data/prices/availability need to be determined. This can be made available to the students through access to the Internet or printed material including brochures, books or booklets that they can use.

Students then work in pairs or groups to solve the problem with the teacher asking suitable questions or prompts and allowing the students to make their decisions. Teachers should ask questions even when it is apparent that students are proceeding well otherwise students may view 'what are you doing?' as a negative question. At the same time students can be guided to reflect on the ways they decide on their approaches and to be encouraged to think critically on whether that approach is productive. Students can develop hypotheses and test and check them as they go through. Most importantly, teachers should encourage students to develop the habit of reflecting critically on their solution methods and outcomes and to consider alternative and potentially better ones.

4.2 *Three examples of the guidelines in use*

The following three problems give examples of how these criteria might be satisfied in practice, and are intended to highlight some of the ways the criteria might be interpreted. They are intended only as a guide to the kinds of problems that might be suitable for encouraging the development of students' reflective thinking. Teachers will be able to think of suitable ones for their own students.

To encourage reflective thinking the teacher can divide the class into small groups of not more than four, so that students can then work on the problem and come up with some hypotheses based for discussion. While the students are working it is good practice for the teacher to use metacognitive questioning to encourage students to reflect on what they have been doing and why they have been doing it.

Each group will record their ideas and steps that they took in the problem solving process, how they made a hypothesis, why it was discarded and how a new one was arrived at. The groups can then share these hypotheses with the other groups as part of a class discussion inviting critique and suggestions from other class members. This would involve discussion about the way that students worked through solving the problem. The teacher is involved in helping the students to understand the mathematical concepts and processes and the metacognitive processes that they went through (Hershkowitz & Schwarz, 1999).

Example 1. Pizza lunch

> You are on your school's Mathematics Olympiad Committee and it is your school's turn to host the inter-school Mathematics Olympiad in Singapore. The committee is required to plan the order of the competitions for the day and you are responsible for planning the refreshments for lunch. The five members of the committee have agreed on pizzas for lunch this year. You have a budget of $250 and 50 competitors are expected to attend. What will you provide as a pizza lunch for the competitors and committee members?

This problem complies with the suggested six characteristics of problem solving suggested above.

(i) Be unfamiliar or non-routine to the student.

It is unlikely that students would have had to do this particular problem before on such a scale. Indeed had they done so previously the teacher might significantly change the character of the problem.

(ii) Have different possible approaches or methods of solution.

A range of approaches can be taken; all need some data collection and analysis. Some students might concentrate on one pizza outlet while others might compare prices and specials across different pizza outlets. Depending on their mathematical knowledge some students may just count the pieces for each size of pizza while others will make comparisons related to the sizes of the pieces.

(iii) Require some thinking about constraints or assumptions.

Assumptions need to be made about the proportions of boys and girls and their appetites, preferred types of pizza crusts and toppings, the proportion of vegetarians, and maximum benefit for lowest cost. This might lead to a random survey of peers to collect and collate data to inform their proposals.

(iv) *Be appropriate to the level of mathematical knowledge and experience of the student.*
This problem includes consideration of numbers of pieces of pizza, the different sizes of pizzas, areas of pizzas and cost per unit area. Thus both number and mensuration are involved using content that is accessible to lower secondary students.

(v) *Be potentially interesting and motivating to the student.*
It is well within the realms of possibility that a student group might be asked to look at such a real life problem. This should present an interesting challenge, especially when there are daily specials and combinations including chocolate puddings!

(vi) *Require justification or explanation of the reasons for the chosen approach and solution.*
There are many potential solutions to providing a pizza lunch so that different groups of students will adopt a number of approaches to this problem. They will need to stop and reflect on their effectiveness both during the process of considering suitable choices and when they have reached their proposed solution.

For this potentially complex problem, similar to the one used in IMPROVE, students need to reflect about what has been done, thinking about how they have organized the data and whether their analysis could be improved. They need to look at costs per person so the size of the pieces and choices of toppings all come under consideration. Most importantly, students need to reflect on the solution they are proposing and justify their choice. This may well lead to a change of thinking about the approach and solution, especially when during presentation and discussion time with other groups of students and their teacher they perceive other solutions to look better than theirs.

Example 2. Filling the jar

> Jon and Lucy play a game called 'Filling the Jar'. They have an unlimited supply of coloured balls and a virtual jar that is defined as full when it contains 30 of the balls. Each player, in turn, puts up to four balls, that is one, two, three or four balls, into the jar; they do not necessarily put the same number of balls in each time. The person to put the 30th ball into the jar is the loser. If you are Lucy and you want to win the game, what should you do? How could this change if the rules were modified?

(i) Be unfamiliar or non-routine to the student.

If the students have already worked out strategies for such a game then the teacher can modify the context. There are many variations on this type of strategy game.

(ii) Have different possible approaches or methods of solution.

Some students will start with the 30 balls and try and work out the solution and end up in a complete muddle. Metacognitive questioning such as 'what are you doing?' or 'can you think of another way?' should help them to reflect on what they are doing and they might perhaps choose a smaller number of balls to fill the jar in the first instance. They might also reflect on their method of data collation and start to record their results in a table and use this to look for patterns. Some students will choose to play out the game with a partner using balls or counters. There are lots of approaches, some more efficient than others!

(iii) Require some thinking about constraints or assumptions.

The students need to think about what happens when they play first or second. How can you work out how to be the winner if you start start first or start second? Is that possible to do? They can be encouraged to think about different rules, such as, what if the last one to play is the winner? How about if you can choose more than four balls and so on?

(iv) *Be appropriate to the level of mathematical knowledge and experience of the students.*
This kind of problem does not assume a high level of mathematical knowledge. It is designed to help students develop hypotheses and to check them. They need to look for patterns, reflect on their methods and see if they can improve on them.

(v) *Be potentially interesting and motivating to the students.*
It is fairly common for students to want to win the games they play, and the playing of a game can in itself be an interesting activity. Although some children's games rely entirely on luck, a game of this kind is interesting because it allows them to play thoughtfully.

(vi) *Require justification or explanation of the reasons for the chosen approach and solution.*
Students will need to justify their proposed strategy for the winning person and explain why the first person can or cannot win. In this case the students may firstly clarify their strategy by playing and winning or losing actual games.

Students need to reflect on their approaches and the mathematics involved. Solutions are not obvious and can take students some time to work them out, change direction, reflect and change again. In this game, for example, it is not uncommon for students to initially think that careful moves are only needed at the closing stage of the game. The discussion between students and students and with the teachers can prompt different ways of thinking and tackling the problem. Recording their approaches and changes in hypotheses helps students to think and to reflect on their processes.

Example 3. Polygons

> If you draw polygonal shapes on 1 cm square dotty grid paper, putting the vertices on the dots, and you are to find the area of each polygon, is it possible to predict the area of the shape by counting the number of dots on the perimeter of the polygon?

(i) *Be unfamiliar or non-routine to the students.*
This problem is probably unfamiliar to many students, although they would be familiar with counting squares and parts of squares to find the area of a polygon.

(ii) *Have different possible approaches or methods of solution.*
They might systematically try all the triangles and then quadrilaterals and so on but they are perhaps more likely to draw random polygons to start with, without any system. Metacognitive questioning should help them to think about, and to reflect upon, their procedures and become increasingly systematic in their choices.

(iii) *Require some thinking about constraints or assumptions.*
Many students will not at first be aware that some polygons they have drawn have no dots inside, while others do have dots inside, although this is not explicitly mentioned in the problem statement.

(iv) *Be appropriate to the level of mathematical knowledge and experience of the students.*
This problem is quite challenging but is within the grasp of secondary one upwards as it involves drawing polygons and working out areas, counting the dots and organizing the data. Once data are carefully organized recognizable patterns are visible.

(v) *Be potentially interesting and motivating to the students.*
It is a bit different and can be fun to draw the shapes on the dotty paper or regular grid paper if dotty paper is not available. The results are a bit surprising too.

(vi) *Require justification or explanation of the reasons for the chosen approach and solution.*

The students will need to justify their results and methods of approach. Reflection on their results and approaches may well lead to new solutions. While secondary students are unlikely to devise formal proofs covering all cases, they can demonstrate the patterns and explain how they can be used to generalize results.

To find solutions students need to draw a range of cases and count the squares and fractions of squares to understand the problem. As they reflect on what they are doing through questioning from the teacher or within their groups they will start to look for patterns and become more systematic. A key step will be recognizing that polygons with various numbers of internal dots should be treated separately. It is best for students themselves to find out rather than being told by their teacher. The reflective learner will make hypotheses and work towards justifying them through discussion and writing about them. They will consider extending the problem perhaps to using triangular rather than square grids, and varying other constraints.

5 Conclusion

It has been illustrated here that problem solving can be an appropriate vehicle for developing and nurturing reflective learners. Teachers can aid the development of students' problem solving processes and reflective learning through the choice of appropriate problem solving activities. For younger or novice problem solvers the problems can be presented with some hints or scaffolding. As the students gain more experience in problem solving the choices that teachers make can be less scaffolded non-routine problems to present students with richer opportunities to develop their reflective learning. To aid teachers in their choice of problems a set of guidelines concerning appropriate criteria has been presented which teachers may use to help them provide problems for students that are particularly useful in developing reflective learners.

Research indicates that collaborative learning in small groups with the use of cognitive questioning can enhance the development of reflective learning. In addition, group and class discussion and justification of solutions with peers and their teacher about their hypotheses is very helpful, as is the written recording of the processes and approaches taken in coming to a solution. Teachers can nurture reflective learners in their mathematics classrooms through engaging students in solving non-routine or open-ended problems and helping them to develop familiarity and expertise with the use of metacognitive questioning.

Acknowlegement

The author thanks colleagues for their feedback on this chapter.

References

Artzt, A.F., Armour-Thomas, E., & Curcio, F.R. (2011). *Becoming a reflective mathematics teacher: A guide for observations and self-assessment* (2nd ed.). New York, NY: Routledge.

Boaler, J. (1993). The role of contexts in the mathematics classroom: Do they make mathematics more real? *For the Learning of Mathematics*, *13*(2), 12-17.

Brophy, J. (1986). Teaching and learning mathematics: Where research should be going. *Journal for Research in Mathematics Education*, *17*, 323-426.

Clarke, D., Waywood, A., & Stephens, M. (1993). Probing the structure of mathematical writing. *Educational Studies in Mathematics*, *25*(3), 235-250.

Clements, M.A., & Ellerton, N.F. (1991). *Polya, Krutetski and the restaurant problem: Reflections on problem solving in school mathematics.* Geelong: Deakin University Press.

Flavell, J. (1976). Metacognitive aspects of problem solving. In L. Resnick (Ed.) *The nature of intelligence* (pp. 231-236). Hillsdale, NJ: Erlbaum.

Flavell, J. (1979). Metacognition and cognitive monitoring: A new area of cognitive-developmental inquiry. *American Psychologist, 34*, 906-911.

Hembree, R. (1992) Experiments and relational studies in problem solving: A meta analysis. *Journal for Research in Mathematics Education, 23*(3), 242-273.

Hershkowitz, R., & Schwarz, B. (1999). Reflective processes in a mathematics classroom with a rich learning environment. *Cognition and Instruction, 17*(1), 65-91.

Kaur, B., & Yeap, B.H. (2009). *Pathways to reasoning and communication in the secondary school mathematics classroom: A resource for teacher by teachers.* Singapore: National Institute of Education.

Kramarski, B. (2008). Promoting teachers' algebraic reasoning and self-regulation with metacognitive guidance. *Metacognition Learning, 3*, 83-89.

Kramarski, B., Mevarech, Z.R., & Arami, M. (2002). The effects of metacognitive instruction on solving authentic tasks. *Educational Studies in Mathematics, 49*(2) 225-250.

Lester, F.K. (1983). Trends and issues in problem solving research. In R. Lesh & M. Landau (Eds), *Acquisition of mathematical concepts and processes* (pp. 229-261). Orlando, FL: Academic Press.

McDonald, J., & Dominguez, L. (2009). Reflective writing. *The Science Teacher, 76*(6), 46-49.

Martinez, M.E. (2006). What is metacognition? *The Phi Delta Kappan, 87*(9), 696-699.

Masingilia, J.O., Davidenko, S., & Prus-Wisnioska, E. (1996). Mathematics learning and practice in and out of school: A framework for connecting these experiences. *Educational Studies in Mathematics, 31*, 175-200.

Masingilia, J.O., & de Silva, R. (2001). Teaching and learning school mathematics by building on students' out-of-school mathematics practice. In B. Atweh, H. Forgasz, & B. Nebres (Eds.), *Sociocultural research on mathematics education* (pp. 329-344). Mahwah, New Jersey: Lawrence Erlbaum Associates, Inc.

Mayer, R.E. (1998). Cognitive, metacognitive, and motivational aspects of problem solving. *Instructional Science, 26*, 49-63.

Mevarech, Z.R. (1999). Effects of meta-cognitive training embedded in cooperative settings on mathematical problem solving. *The Journal of Education Research, 92*, 195-205.

Mevarech, Z.R., & Amrany, C. (2008). Immediate and delayed effects of metacognitive instruction on regulation of cognition and mathematics achievement. *Metacognition Learning, 3*, 147-157.

Mevarech, Z.R., & Kramarski, B. (1997). IMPROVE: a multidimensional method for teaching mathematics in heterogeneous classrooms. *American Educational Research Journal, 34*, 365-394.

Mewborn, D.S. (1999). Reflective thinking among pre-service elementary teachers. *Journal for Research in Mathematics Education, 30*, 316-341.

Ministry of Education (2006). *Mathematics syllabus – Secondary.* Singapore: Author.

Moon, J. (2004). *A handbook of reflective and experiential learning: Theory and practice.* London: RoutledgeFarmer.

National Council of Mathematics Teachers (NCTM). (2000). *Principles and standards for school mathematics.* Reston, VA. Author.

Polya, G. (1990). *How to solve it.* Reading, Berks: Cox and Wyman Ltd.

Rodgers, C. (2002). Defining reflection: Another look at John Dewey and reflective thinking. *Teachers College Record, 104*(4), 842-866.

Schneider, W., & Artelt, C. (2010). Metacognition and mathematics education. *ZDM – The International Journal on Mathematics Education*, 42, 149-161.

Schoenfeld, A.H. (1985). *Mathematical problem solving.* Orlando, FL: Academic Press.

Schoenfeld, A.H. (1987). What's all that fuss about metacognition? In A.H. Schoenfeld (Ed.), *Cognitive science and mathematics education* (pp. 189-215). Hillsdale, NJ: Erlbaum.

Schoenfeld, A.H. (1992). Learning to think mathematically: Problem Solving, metacognition, and sense making in mathematics. In D. Grouws (Ed.) *Handbook for research on mathematics teaching and learning* (pp. 334-370) New York: MacMillan.

Schoenfeld, A.H. (2011). *How we think: A theory of goal-orientated decision making and its educational applications.* New York: Routledge.

Shimizu, Y. (2009). Japanese approach to teaching mathematics via problem solving. In B. Kaur, Yeap, B.H., & Kapur, M. (Eds). *Mathematical problem solving: Yearbook 2009* (pp. 89-101). Singapore: World Scientific Publishing Co. Pte. Ltd.

Stanic, G., & Kilpatrick, J. (1989). Historical perspectives on problem solving in the mathematics curriculum. In R. Charles & E. Silver (Eds.) *The teaching and assessing of mathematical problem solving* (pp. 1-22). Reston, VA: National Council of Teachers of Mathematics.

Toh, T.L., Quek, K.S., Leong, Y.H., Dindyal, J., & Tay, E.G. (2011). *Making mathematics practical: An approach to Problem Solving.* Singapore: World Scientific Publishing Co. Pte. Ltd.

Wheatley, G.H. (1992). The role of reflection in mathematics learning. *Educational Studies in Mathematics, 23*(5), 529-541.

Zevenbergen, R. (1997). Situated numeracy: A case study of pool builders. In N. Scott & H. Hollingsworth (Eds.), Proceedings of the 16th Biennial Conference of the Australian Association of Mathematics Teachers, *Mathematics creating the future* (pp. 89-94). Melbourne: AAMT.

Chapter 10

Mathematics Competition Questions and Mathematical Tasks for Instructional Use

TOH Tin Lam

Mathematics competition questions are useful resources for teachers to craft mathematical tasks for instructional purposes. These tasks could serve two objectives: to develop (i) students' interest in mathematics and (ii) higher order thinking skills in the subject. The learning experiences provided by these tasks are rich and certainly appropriate for nurturing reflective learners. This chapter illustrates through examples how mathematics competition questions may be crafted into useful tasks for use in classrooms to advance the above two objectives.

1 Introduction

Education and competition are intimately related. It is natural for children to compete with one another; hence it is not surprising that some educators are of the opinion that competition can be put to educational use. However, there is no consensus among education theorists on whether competitions should be encouraged or constrained (Verhoeff, 1997).

Most mathematics educators and mathematicians identify the key role of mathematics competitions as developing the mathematically gifted (see, Bicknell, 2008; Campbell & Walberg, 2010; Kalman, 2002). For example, many mathematicians are generally of the opinion like Xu (2010), who identifies the role of mathematics competition to "improve the mathematical thinking and technical ability in solving mathematical problems" (p. v) for the higher ability students. Through various forms of mathematics competition, researchers have also identified the broad possibility for cooperation across mathematics and other disciplines at the international level (see, France & Andzans, 2008). Focus on developing the education of the higher ability or mathematically talented students abounds in the literature on the roles of mathematics competitions. However, few studies have been done on how mathematics competitions can contribute in the mathematics classrooms for the benefit of the general student population.

Many of the mathematics competition questions have rich educational values other than that for the mathematically talented students (Toh, 2012). This chapter discusses how many of the mathematically rich competition questions can serve as useful resources for teachers to craft mathematical tasks for instructional purposes in mathematics classrooms for students using sound pedagogical principles. These tasks could serve to develop students' interest in school mathematics and also higher order thinking skills in the subject.

Illustrations used in this chapter are taken from the past year questions from both the Junior (students age 13 to 14) and Senior (students age 15 to 16) teams of the Singapore Mathematical Olympiad (SMO), which is the most prestigious mathematics competition in Singapore at the national level. Moreover, interested readers will have easy access to these materials through the publications of the Singapore Mathematical Society or their website.

2 Mathematical Tasks and Competition Questions

Mathematics competition apparently challenges the traditional attitude towards mathematics, in which the subject is seen as one whose main business involves "extensive drill-and-practice within a formulaic environment" (Crawford & Brown, 2002). An examination of the mathematics competition questions clearly suggests that a good problem solver in the mathematics competitions is not produced solely through extensive drill-and-practice in the algorithmic procedures and mindless application of formulae; sound mathematical reasoning and clear thinking skills are more crucial for the problem solver.

On the other hand, a careful study of the competition questions also reveals that many of these competition questions are not beyond the grasp of an average student. Many of the questions are based on sound pedagogical principles which are not usually found in the school textbooks. In this way, the competition questions form a repertoire of questions which are mathematically "rich" and could complement the school textbooks as useful resources for teaching. These questions could provide meaningful learning experiences for students learning mathematics. In particular, these questions can serve to engage students in Higher Order Thinking Skills in mathematics classrooms (Toh, 2012).

In their original form as they appear in the mathematics competitions, these questions might appear to be beyond the grasp of an average student. In this chapter, we shall demonstrate how effective mathematical tasks with sound pedagogical principles can be crafted from some of these questions.

A mathematical task is defined as a set of problems or a single complex problem that focuses students' attention on a particular mathematical idea (Stein, Grover, & Henningsen, 1996). International research studies have identified that students spend over 80% of the curriculum time in mathematics class work on tasks (Kaur & Toh, 2012). According to Doyle (1988), "the work students do, defined in large measures by the tasks teachers assign, determines how they think about a curricular domain and come to understand its meaning" (p. 167). Hence, different kinds of tasks lead to different types of instruction, which subsequently lead to different learning opportunities for students (Doyle, 1988; Stein, Smith, Henningsen, & Silver, 2000).

Mathematical tasks crafted from mathematics competition questions provide students with rich learning experiences that can serve to nurture the general student population into reflective learners in two ways: firstly, the tasks could be used to develop students' interest in the subject. Secondly, the learning experience acquired through these activities develops students' higher order thinking skills in mathematics. In this way, students could even be challenged to re-examine their originally acquired mathematical notions and beliefs.

3 Developing Students' Interest in Mathematics

For the convenience of demonstrating how rich mathematical tasks can be crafted from mathematics competition questions, we shall restrict our discussion to questions from the algebra strand throughout this chapter. In learning algebra, it is generally proposed that students be given a formal introduction with real-life situations which promote some fundamental algebraic ideas (Ursini, 2001; Yeap, 2009). It is also important for students to appreciate the link between algebra and arithmetic through their appreciation of the application of algebra in performing arithmetic computation. This is not normally emphasized in many mathematics classrooms (Toh, 2011).

3.1 *Illustration 1: Compute using algebraic formulae*

Consider the following group of past year competition questions on elementary algebra obtained from the Singapore Mathematical Olympiad (SMO) for both junior (Age 13 and 14) and senior categories (Age 15 and 16). We could identify one core pedagogical principle that underlies all the questions in Figure 1.

The rather cumbersome numerical expressions could be evaluated using a scientific calculator, but keying these long chains of numbers into the calculator would be rather tedious. However, if algebraic identities (such as the algebraic identities for the perfect squares, $(a \pm b)^2 = a^2 + b^2 \pm 2ab$, or that for the difference of two squares, $(a+b)(a-b) = a^2 - b^2$) are used, the answers to the questions in Figure 1 could be easily obtained without the use of a calculator. The group of questions in Figure 1 brings out the extremely important principle that elementary algebraic identities frequently help to make arithmetical computations easy. This message is particularly important today in this age of advanced instructional technology – one should not rely exclusively on the use of technology. Mathematical tasks could be crafted based on the ideas provided by these competition questions as demonstrated in Task 1 (see Figure 2).

- Find the value of $\dfrac{2011^2+2111^2-2\times2011\times2111}{25}$.

  (SMO 2011 Senior Q11)

- Find the value of $\dfrac{2007^2+2008^2-1993^2-1992^2}{4}$.

  (SMO 2007 Junior Q13)

- Multiple choice question:

  What is the value of $\dfrac{2005^2+2\times2005\times1995+1995^2}{800}$?

  (A) 20000 (B) 2000 (C) 200000
  (D) 2000000 (E) None of the above
  (SMO 2005 Senior Q2)

Figure 1. A selection of past year competition questions on elementary algebra

If students are to learn something, they must first be willing to learn it. Motivation is an important factor in the success of students' learning activity. Task 1 makes the learning of algebraic formulae meaningful and no longer a chore, as it becomes clear that algebraic formulae are much more efficient than even the most advanced scientific calculators. It becomes very clear to students that many seemingly complicated expressions can be calculated much more quickly than calculating tools.

Task 1.

Without using your calculator, write down the answers of the following expressions. Which rules have you used? Discuss with your partner beside you.

- $(2009)^2 - 2(2009)(2008) + (2008)^2 =$ __________
- $(2012)^2 - 2(2012)(2011) + (2011)^2 =$ __________
- $(79)^2 + 2(79)(21) + (21)^2 =$ __________
- $(0.541)^2 + 2(0.541)(0.459) + (0.459)^2 =$ __________
- $(1234)^2 - (1233)^2 =$ __________
- $(3211)^2 - (3210)^2 =$ __________
- $(45)^3 - 3(45)^2(44) + 3(45)(44)^2 - (44)^3 =$ __________
- $(0.5)^3 + 3(0.5)^2(-1.5) + 3(0.5)(-1.5)^2 + (-1.5)^3 =$ ______

More challenging questions:

- $\dfrac{(2015)^2 - 2(2015)(2013) + (2013)^2}{(1035)^2 - 2(1035)(1033) + (1033)^2} =$ __________
- $\dfrac{(10235)^2 - 2(10235)(10234) + (10234)^2}{(12)^3 - 3(12)^2(11) + 3(12)(11)^2 - (11)^3} =$ __________
- $\dfrac{2011^2 + 2111^2 - 2 \times 2011 \times 2111}{25} =$ __________
- $\dfrac{2005^2 + 2 \times 2005 \times 1995 + 1995^2}{800} =$ __________
- $\dfrac{2007^2 + 2008^2 - 1993^2 - 1992^2}{4} =$ __________

Make a note of the rules of the algebra that you have used to answer the above questions in the space provided below:

Rule 1: __________
Rule 2: __________
Rule 3: __________
Rule 4: __________

Figure 2. Sample of an algebra task that is based on competition questions in Figure 1

3.2 *Illustration 2. Compute using basic algebraic / arithmetic rules*

Consider another set of competition questions on arithmetic computation.

- Find the value of 1998 x 19971997 – 1997 x 19981998. (SMO 1998 Junior Q10)
- Find the value of 123 x 567567 – 567 x 123123. (SMO 2002 Junior Q10)

Figure 3. A selection of past year competition questions on arithmetic computation

It is clear that both questions in Figure 3 yield the same value (zero). As an illustration, in the first question of Figure 3, it could be seen that

$$\begin{aligned}19971997 &= 19970000 + 1997\\ &= 1997\ (10000 + 1)\\ &= 1997 \times 10001\end{aligned}$$

Similarly,

$$19981998 = 1998 \times 10001$$

(Alternatively, we could also see from division that 19971997 = 1997 x 10001) It then becomes clear that 1998 x 19971997 = 1998 x 1997 x 10001 while 1997 x 19981998 = 1997 x 1998 x 1001, hence their difference equals zero. This set of questions shows the "concrete" application of either the distributive law of multiplication over addition in arithmetical computation or division of a large numbers with a repeating unit. The numerical expressions of Figure 3, unlike that in Figure 1, are not tedious to evaluate using a calculator. However, teachers could invite their students to observe that the final answers for both questions are zero, and challenge them to explain the reason. This is one approach of arousing students' curiosity — a crucial first step to problem solving (Toh, 2009).

Task 2 in Figure 4 reinforces the underlying principle behind the questions in Figure 3. Here scaffolding is provided as hints to students to facilitate them relate the computations with the distributive law through its application.

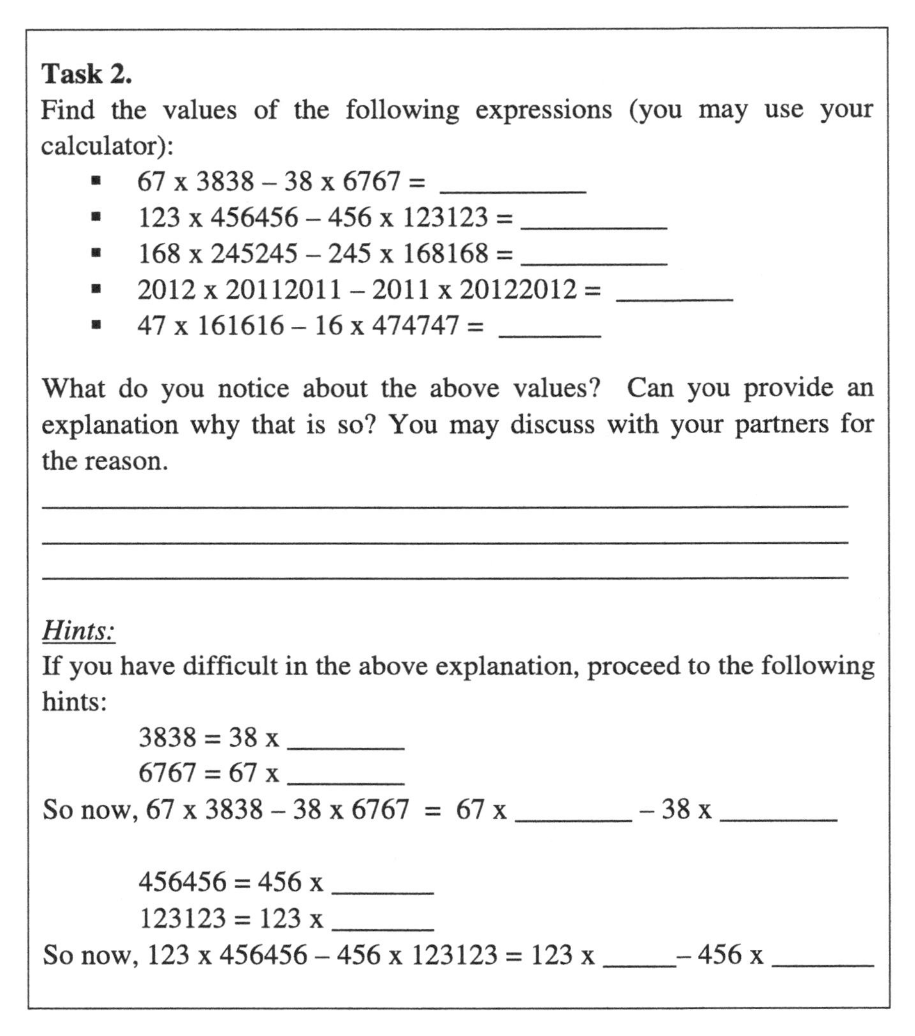

Task 2.

Find the values of the following expressions (you may use your calculator):

- 67 x 3838 – 38 x 6767 = __________
- 123 x 456456 – 456 x 123123 = __________
- 168 x 245245 – 245 x 168168 = __________
- 2012 x 20112011 – 2011 x 20122012 = ________
- 47 x 161616 – 16 x 474747 = _______

What do you notice about the above values? Can you provide an explanation why that is so? You may discuss with your partners for the reason.

____

____

____

Hints:

If you have difficult in the above explanation, proceed to the following hints:

3838 = 38 x ________

6767 = 67 x ________

So now, 67 x 3838 – 38 x 6767 = 67 x ________ – 38 x ________

456456 = 456 x _______

123123 = 123 x _______

So now, 123 x 456456 – 456 x 123123 = 123 x _____– 456 x _______

Figure 4. Sample of a task that is based on competition questions in Figure 3

3.3 *Illustration 3. Game based on a competition question*

Mathematical games can be useful tools to enthuse students to learn the school subject. Games provide opportunities for students at all levels to engage in constructing strategies, formulating hypothesis, asking new questions, and even posing new problems (Pinter, 2011). Figure 5 shows an example of a competition question that can potentially be developed into mathematical games for use in mathematics classrooms.

<u>Multiple Choice Question</u>
The numbers 1, 2, 3,…., 2003 are written on a board. An operation is performed as follows: two numbers x and y on the board are erased and replaced by $x + y - 1$. This operation is repeated until only one number N is left on the board. What is N?

(A) 2003002 (B) 2005004
(C) 2007006 (D) 2009008
(E) None of the above
(SMO 2003 Senior Q7)

Figure 5. A competition question that has a rich context for a game

The mathematical ideas underlying the question in Figure 5 include the following:

- addition of numbers is both commutative and associative, i.e., the order of addition and performing the binary operation of addition does not matter;
- the formula for the sum of the first n positive integers $1+2+3+....+n=\frac{n(n+1)}{2}$.

The question in Figure 5 provides a creative way to engage students into thinking about the commutativity and associativity of the addition process: "Does it matter on the order that two numbers are erased? Can we switch the order of additions when two or more numbers are added? Why or why not?" It is obvious that the sum of the first 2003 positive integers might not be manageable for an average student. To start with, one might modify the question into a series of manageable activities (with smaller numbers and simpler rules in the game so that the emphasis on applying the formula for the sum of the first n positive integers can be downplayed) for use in a mathematics classroom when one approaches it as follows:

Task 3.

In this lesson we are going to play a game in pairs.

Activity 1

Write down the numbers 1,2,3,4 and 5 on the whiteboard.

Erase off any two numbers x and y, then replace by the sum $(x+y)$ of the two numbers erased on the board.

Repeat this process until only one number is left on the board. What is this number that is left?

Play this game again by first erasing off another two numbers. What is the last number that remains on the board?

- o How many times do you need to erase until you get only one number on the board?
- o What is the last number for the above two games?
- o Are the last number the same for the above two trials? Can you explain why?

Activity 2

Write the numbers 1,2,3,….., 10 on the board. Repeat Activity 1.

- o Make a prediction what is the last number that remains on the board. Can you explain why?
- o How many times do you think you need to erase until you get only one number left on the board?
- o Try out the process to verify your prediction is correct.
- o Do you think that the order of erasing sets of two numbers affect the last number on the board? Why?

Activity 3

Now that you have seen the pattern above, make a prediction to the game below.

- o The numbers 1, 2, 3,…., 10 are written on the board. Each time two numbers x and y are erased, it is replaced by the new number $x + y - 1$. What is the last number left on the board?

Figure 6. Sample of a game that is based on a competition question in Figure 5

What makes each of the Activities in Task 3 exciting for the students is that regardless of the order of how pairs of numbers are erased, the final number appearing on the board is unchanged. Students would be amused and thereby encouraged to explore the reason for this.

4 Developing Students' Higher Order Thinking

The traditional attitude towards mathematics among school students is usually one that emphasizes extensive drill-and-practice within a formulaic environment (Crawford & Brown, 2002). As students' learning environment in particular and the world in general are changing rapidly, educational expectations are also shifting. Emphasis on higher order thinking skills should be of greater concern within the school mathematics lessons than mere drill-and-practice or acquisition of pieces of knowledge. What is higher order thinking then? According to Thomas, Thorne, and Small (2001),

> Higher Order Thinking, or HOT for short, takes thinking to higher levels than just restating the facts. HOT requires that we do something with the facts. We must understand them, connect them to each other, categorize them, manipulate them, put them together in new or novel ways, and apply them as we seek new solutions to new problems (quoted by Crawford & Brown, 2002, p. 7).

For example, getting students to challenge their thinking or beliefs about their understanding of mathematical procedures or concepts is one way to engage them in higher order thinking skills.

4.1 *Challenging students' understanding of mathematical steps to solve an algebraic equation*

Consider the following SMO problem which test students' ability to find the solution of an algebraic equation.

Multiple Choice Question

How many real numbers x satisfy the equation

$$\frac{x^2-x-6}{x^2-7x-1}=\frac{x^2-x-6}{2x^2+x+15}?$$

(A) 4 (B) 3 (C) 2 (D) 1 (E) 0

(SMO 2004 Senior Q7)

Figure 7. A competition question on solving algebraic equations

The question in Figure 7 addresses one very common misconception, i.e. cancelling identical terms from both sides of a given equation, among students when solving algebraic equations. What makes this question challenging to most students is that both sides of the equation involve algebraic fractions instead of polynomial functions. Teachers could craft the following task, shown in Figure 8, based on the spirit of the question to engage students in examining their conception of solving algebraic equations.

Task 4.

Joseph and Pearlyn were asked to solve $\frac{x^2-x-6}{x^2-7x-1}=\frac{x^2-x-6}{2x^2+x+15}$.

Joseph

Identifying that there is a common term on both sides of the equation, Joseph cancelled off the common term and obtains

$$\frac{1}{x^2-7x-1}=\frac{1}{2x^2+x+15}$$

Hence $2x^2+x+15=x^2-7x-1$, which reduces to $x^2+8x+16=0$. By factorizing the quadratic expression, Joseph obtained $x = -4$.

Pearlyn

Instead of cancelling off the like terms from both sides of the equation, Pearlyn performed cross-multiplication to obtain

$$(2x^2+x+15)(x^2-x-6)=(x^2-7x-1)(x^2-x-6).$$

Still, she did not cancel off the common term, but perform as follows:

$$(2x^2+x+15)(x^2-x-6)-(x^2-7x-1)(x^2-x-6)=0$$
$$(x^2-x-6)[(2x^2+x+15)-(x^2-7x-1)]=0$$
$$(x^2-x-6)(x^2+8x+16)=0$$

By completely factorizing both quadratic expressions, Pearlyn obtained the solutions $x = 3, -2, -4$.

Question 1: Can you explain what has happened? Who is correct and who makes an error?

Question 2: What have you learnt from the above episode?

Try:
By using any correct method, solve the following equation:

- $(x^2-1)(2x^2-5x-8)=(x^2-1)(x^2+x+1)$.

Check with your partner and compare the method that your partner has used.

Figure 8. A task challenging students' conception of solving an algebraic equation

4.2 *Challenging students' understanding of mathematical reasoning*

While pattern recognition is an important part of the lower secondary school mathematics curriculum, it is also important to provide opportunity for students to be engaged in activities demonstrating that inductive reasoning by pattern recognition is not infallible, or "pattern is not proof" (e.g. Toh, Quek, Leong, Dindyal, & Tay, 2011).

A sample of the past year competition question meant for the lower secondary school students is shown in Figure 9.

Multiple Choice Question

Seven points lie on the circumference of a circle. To every two points a chord is drawn. What is the maximum number of regions that these chords can divide the circle?

(A) 31 (B) 32 (C) 51 (D) 57 (E) 64

(Source: SMO Junior, original source unavailable)

Figure 9. Sample of a question that demonstrates that pattern is not proof

A student using pattern recognition in solving the above question will conclude E as the answer (whereas D is the correct answer). In this question, the pattern formed by the fewer number of points on the circle will suggest an incorrect number pattern with increasing number of points. It is *not* crucial to teach the students the mathematics underlying this question (which is a sophisticated topic for secondary school students), but it is useful to introduce to students as part of their learning experience in school mathematics that pattern recognition is not infallible. Figure 10 shows a sample task can be crafted to engage students in this invaluable learning experience.

Task 5.

You are given a circle and n points on the circumference of the circle. To every pair of these n points, construct a chord passing through. Ensure that the number of regions divided by these chords must be the maximum (e.g. no two chords must be parallel, no three chords concurrent). Let T_n denote the maximum number of regions divided by the chords formed by the n points. The first few have been done for you.

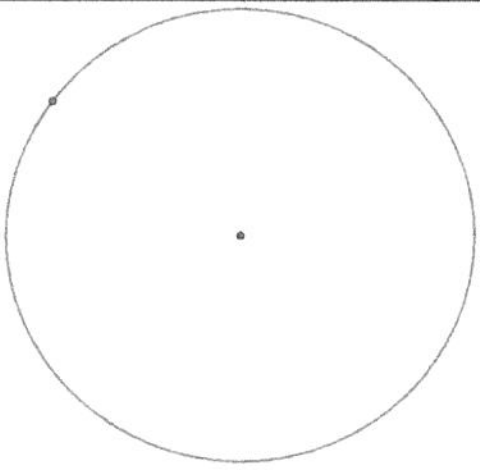	$T_1 = 1$
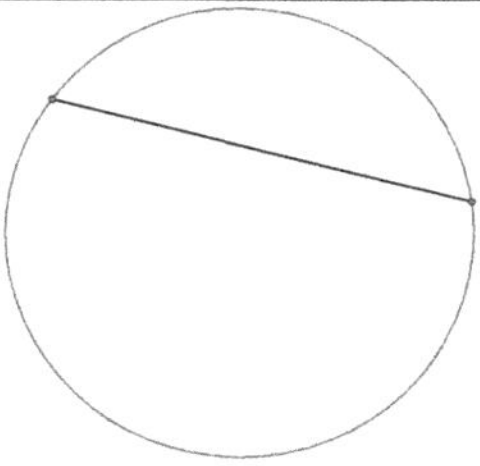	$T_2 = 2$
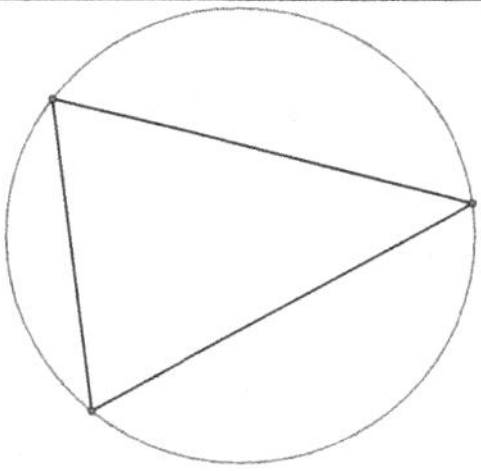	$T_3 = 4$
	$T_4 =$
	$T_5 =$

- Make a prediction of the formula for T_n.
- By actual drawing of the diagram, find the values of T_6 and T_7. What do you observe?
- What have you learnt from the above activity?

Figure 10. A task challenging students' understanding of pattern recognition

The actual formula for T_n, is mathematically difficult for secondary school students but could be a useful task for higher level students. This activity would also serve as a good introductory activity at the Pre-university level for students to appreciate the importance of learning higher level mathematics (e.g. Toh, 2006).

5 Some Valuable "Obsolete" Questions in Mathematics Competitions

With the evolving school mathematics curriculum and related pedagogy for classroom teaching, some topics within the normal mathematics curriculum are rendered "obsolete" under the new educational setting, especially tasks that can easily be completed using technology. However, the techniques required to complete these tasks could yield creative use of the standard mathematical algorithms or deeper understanding of school mathematical knowledge. The set of questions in Figure 11 serves as a good illustration of this type of questions.

Question 1: Suppose $a = \sqrt{6} - 2$ and $b = 2\sqrt{2} - \sqrt{6}$. Then

(A) $a > b$ (B) $a = b$ (C) $a < b$

(D) $b = \sqrt{2}\,a$ (E) $a = \sqrt{2}\,b$

Question 2: Simplify $144\left(\sqrt{7+4\sqrt{3}} + \sqrt{7-4\sqrt{3}}\right)$.

Question 3: Given that $2\sqrt{x} - \sqrt{4x-11} = 1$, find the value of the expression $x^2 + x + 1$.

(A) 71 (B) 81 (C) 91 (D) 47 (E) 63

Figure 11. Sample of three SMO questions which are considered "obsolete"

As stipulated in the secondary school mathematics curriculum, calculators are allowed at all school examinations. Thus, Questions 1 and 2 would become meaningless calculator pressing exercises. However, without the use of calculators, these two questions require the creative use of algebraic manipulation. In Question 1, the solver is required to irrationalize the denominator of each of the terms a and b (while in the usual mathematics curriculum, students are generally expected to rationalize the denominators of surdic fractions). This provides students with opportunity to examine both methods critically in appropriate situations.

Question 2 requires the solver to find the square of the surdic expression and then take the square root again. This is obviously another creative use of the algebraic techniques not usually emphasized in the school mathematics curriculum. Question 3 involves the solving of a surdic equation. In solving such an equation, the solver is expected to perform squaring twice to get rid of the square roots and creatively isolate the surdic expression that arises from the squaring process.

6 Some Final Remarks

This chapter clearly demonstrates with illustrations that mathematics competition questions could serve a much wider function beyond useful tools for identifying the mathematically talented. Competition questions can serve as a rich resource for teachers to craft instructional tasks for their daily classroom needs. As illustrated above, the tasks developed could serve as motivation for students to learn mathematics or to challenge them in higher order thinking skills. Teachers in crafting instructional tasks from competition questions should "maintain the cognitive demand of the tasks at their intended level" and not "proceduralize the tasks and place emphasis on students' remembering and applying rules and formulas" (Hill, Ball, & Schilling, 2008).

References

Bicknell, B. (2008). Gifted students and the role of mathematics competitions. *Australian Primary Mathematics Classroom, 13*(4), 16-20.

Campbell, J.R., & Walberg, H.J. (2010). Olympiad studies: Competitions provide alternatives to developing talents that serve national interests. *Roeper Review, 33*(1), 8-17.

Crawford, C.M., & Brown, E. (2002). *Focusing upon higher order thinking skills: Webquests and the learner-centred mathematical learning environment.* Retrieved from ERIC Database.

Doyle, W. (1988). Work in mathematics classes: The context of students' thinking during instruction. *Educational Psychologist, 23*, 167-180.

France, I., & Andzans, A. (2008). *How did the prodigal son save his skin.* Paper from ICME11 Discussion Group 19: The role of mathematical competition and other challenging contexts in the teaching and learning of mathematics.

Hill, H., Ball, D., & Schilling, S. (2008). Unpacking pedagogical content knowledge: Conceptualising and measuring teachers' topic-specific knowledge of students. *Journal for Research in Mathematics Education, 39*(4), 372-400.

Kalman, R. (2002). Challenging gifted students: The math Olympiads. *Understanding Our Gifted, 14*(4), 13-14.

Kaur, B., & Toh, T.L. (2012). Reasoning, communication and connections in mathematics: An introduction. In B. Kaur & T.L. Toh (Eds.), *Reasoning, communication and connections in mathematics: AME yearbook 2012* (pp. 1-10). Singapore: World Scientific Publishing.

Pinter, K. (2011). Creating games from mathematical problems. *PRIMUS, 21*(1), 73-90.

Stein, M.K., Grover, B., & Henningsen, M. (1996). Building student capacity for mathematical thinking and reasoning: An analysis of mathematical tasks used in reform classrooms. *American Educational Research Journal, 33*, 455-488.

Stein, M.K., Smith, M.S., Henningsen, M., & Silver, E.A. (2000). *Implementing standards-based mathematics instruction: A casebook for professional development.* New York: Teachers College Press.

Toh T.L. (2006). Mathematical reasoning from O-level to A-level. *Mathematical Medley, 33*(2), 34 - 40.

Toh, T.L. (2009). Arousing students' curiosity and mathematical problem solving. In B. Kaur & B.H. Yeap (Ed.), *Mathematical problem solving* (pp. 241-262). Singapore: World Scientific.

Toh, T.L. (2011). Exploring mathematics beyond school curriculum. In L.A. Bragg (Ed.) *Maths is multi-dimensional* (pp. 77-86). Melbourne, Australia: Mathematical Association of Victoria.

Toh, T.L. (2012). *The roles of mathematics competition in Singapore mathematics education*. Paper presented at ICME12, July 2012, Seoul.

Toh, T.L., Quek, K.S., Leong, Y.H., Dindyal, J., & Tay, E.G. (2011). *Making mathematics practical: An approach to problem solving*. Singapore: World Scientific.

Ursini, S. (2001). General methods: A way of entering the world of algebra. In R. Sutherland, T. Rojano, A. Bell, & R. Lins (Eds.) *Perspectives on school algebra* (pp. 209-230). Dordrecht: Kluwer Academic Publishers.

Verhoeff, T. (1997). *The role of competition in education*. Paper presented at Future World: Educating for the 21st Century, a conference and exhibition at IOI'97 December 1997. Retrieved September 17, 2011, from http://www.win.tue.nl/~wstomv/publications/competit.pdf

Xu, J.G. (2010). *Lecture notes on Mathematical Olympiad courses: For Junior section volume 1*. Singapore: World Scientific.

Yeap, B.H. (2009). Teaching of algebra. In P.Y. Lee & N.H. Lee. (Eds.) *Teaching secondary school mathematics: A resource book* (pp. 25-50). Singapore: McGraw-Hill.

Chapter 11

Mathematical Tasks in the Advanced Mathematics Class for Nurturing Reflection

Oh Nam KWON Ji Eun LEE

Significant changes were made to the Korean mathematics curriculum in 2009. The revised high school curriculum is composed of elective subjects. Advanced mathematics is an elective subject introduced for the first time in the high school curriculum. Therefore it is posing a challenge to the teachers who are teaching it. At the advanced level, it is important for learners to reflect on their mental mathematical activities. This chapter defines reflective thinking as looking back on one's problem solving and thinking processes. It analyzes tasks that may nurture students' reflective thinking specifically for the teaching and learning of polar coordinates.

1 Introduction

1.1 *Advanced mathematics — A new challenge for the mathematics teacher in Korea*

Korea has a national curriculum system. Korean mathematics curricula have been revised in response to rapidly changing external environment in recent decades. There has been the government announcement about major curriculum revisions in 1997, 2007, and 2009. The latest one, called 'creativity and character building curriculum', was constructed to reflect the national and social demand that Korea should cultivate more creative and talented people for the globalization era. The latest mathematics curriculum, revised in 2009, emphasizes mathematical creativity designed to equip students with capacities on basic learning ability, divergent thinking ability, problem-solving ability, originality, and ability to create new values. It also reinforces mathematical process including problem-solving ability, reasoning ability, and communication ability (Ministry of Education, Science and Technology, 2011).

A significant change in the 2009 revised mathematics curriculum is that the high school curriculum comprises electives and allows students to choose subjects which suit their aptitude, future career and goals. The mathematics curriculum is divided into core subjects, basic subjects, and advanced subjects, thus allowing students to choose subjects according to their levels and complete the curriculum. Table 1 shows the elective subjects in the mathematics curriculum. Tables 2 and 3 show the main syllabus of the advanced mathematics I and II courses respectively.

Table 1
Elective subjects in the 2009 revised mathematics curriculum

Basic course	**Core course**		**Advanced course**
Basic mathematics	Mathematics Ⅰ	Calculus Ⅱ	Advanced mathematics Ⅰ
	Mathematics Ⅱ	Geometry and Vectors	Advanced mathematics Ⅱ
	Calculus Ⅰ	Probability and Statistics	

Table 2
Advanced mathematics I

Area	Content
Vectors and matrix	vectors
	matrix and simultaneous linear equations
Linear transformation	linear transformation and matrix
	eigenvalues and multiplication of matrix
Graphs	definition of graphs
	various graphs
	applications of graphs

Table 3
Advanced mathematics II

Area	Content
Complex numbers and polar coordinates	polar form of complex numbers
	polar coordinates and polar equations
Application of calculus	applications of differentiation
	differential equations
	applications of integration
Partial differentiation	definition of two variable function
	limit and continuity
	partial differentiation
	application of partial differentiation

Students who are good at and interested in mathematics can take and complete 'Advanced mathematics Ⅰ', 'Advanced mathematics Ⅱ'. This allows them to study mathematics more deeply.

Advanced mathematics is a kind of advanced placement course which is an intense course of the existing educational program and deals with new contents. As advanced mathematics has been introduced for the first time in high school curriculum, teachers of advanced mathematics need to become familiar with their materials and mathematical tasks that support the latest emphasis of the curriculum. In this chapter, feasible mathematical tasks that may help teachers nurture reflective thinking in their classrooms are introduced and analyzed.

1.2 *The necessity and importance of learning polar coordinates*

The necessity and importance of polar coordinates in advanced mathematics class can be explained by its application to other disciplines such as engineering and physics. First, many physical systems are simpler and more intuitive to model using polar coordinates. The initial motivation for the introduction of the polar system was the study of circular and orbital motion (Boyer, 1949). Coordinates are shown in figure algebraic expressions making it possible to convert geometric problems into algebraic problems. As a result, the concept of calculus was born and the discipline of mathematics grew explosively. However, many students do not realize the importance of coordinates. Through experiencing polar coordinates instead of rectangular coordinates, which students are familiar with, students will be able to engage in reflective thinking about the necessity and importance of different coordinate systems. Also, students will realize that there are many ways of showing the location of a point, thereby improving flexibility of thinking.

This chapter focuses on the topic polar coordinates and explores several tasks that may be used by teachers to nurture reflective thinking among their students.

2 Reflective Thinking and Reflective Learning

In this section, we review some relevant literature and outline the stages of facilitating reflective learning and characteristics of tasks that nurture reflection.

2.1 *Reflective thinking*

The importance of reflective thinking in the education of thinking was emphasized by John Dewey. Dewey (1933) described that reflection is a meaning making process that moves a learner from one experience into the next with deeper understanding of its relationships. Piaget's (1976) construct of "reflexive abstraction," for example, is a mechanism for extracting, reorganizing, and consolidating knowledge. He asserted that especially mathematical teaching should require students to make conscious reflection on such structure.

Researchers of mathematical education including Polya, Freudenthal, and Skemp also emphasized the importance of the role of reflective thinking in nurturing the ability of mathematical thinking. In particular, Polya (1973) mentioned the importance of reflection in mathematics problem solving. Polya highlights four main phases of the problem-solving process: understanding the problem, devising a plan, carrying out the plan and looking back at the completed solution. The final stage requires students to be reflective in determining which part of their solution worked, which part did not, why it did not work and what strategy could be used to solve other similar problems in future. Skemp (1979) distinguishes between two levels of functioning of intelligence: the intuitive and the reflective. He speaks of "reflective intelligence" as the "ability to make one's own mental processes the object of conscious observation" (p. 175). Freudenthal (1978, 2002), too, emphasized that much of mathematics is about reflecting on the activities of oneself and others, and reflection is necessary for mathematization at higher levels. Students were recommended to acquire an attitude of reflecting on their activities for successful learning of mathematics.

In summary, though the term reflection is defined and used slightly differently among scholars, in common it is being aware of one's own activities or thoughts or uses them again as objects of thinking. In doing so, it is possible to have an insight of the contents and reach the advanced level. Therefore, in this chapter, reflective thinking is regarded as looking back on one's problem solving and thinking process. Next we will inquire into the ways used to facilitate reflective learning.

2.2 *Cognitive disequilibrium for reflective thinking*

It is not guaranteed that students will have reflective thinking voluntarily in mathematical learning. Ge and Land (2003) found that students did not voluntarily construct arguments to justify their proposed solutions without being prompted to do so. Therefore, the teacher may make proper interventions to help students do reflective thinking consciously.

The cognitive development theory for this can be found in Piaget's theory of equilibration. The equilibration (Piaget, 1967, 1977) means that the results of the organism's actions feed back to its psychological structure in a way that is discrepant with that structure. This discrepancy induces a state of disequilibrium, leading to reorganization of the existing structure in a way that serves to restore equilibrium. The attempt is made to artificially induce disequilibrium by presenting models that reflect a structure's discrepant from the students'. Insofar as the students become aware of the discrepancy between the structure of the model and their own structure, this discrepancy might induce the disequilibrium leading to reorganization. Briefly, the mechanisms of equilibration in the individual are triggered by the disturbance of a knowledge system when a 'gap' or conflict is recognized by matching anticipation, generated by the application of it.

If the teacher can create disequilibrium intentionally for a student, the student will feel the necessity for re-equilibration. Reflective thinking occurs as a cognitive action for the understanding of disequilibrium. In the following sub-section we will discuss how teachers may create cognitive disequilibrium for students.

2.3 *Mathematical tasks for cognitive disequilibrium*

The teacher may use questions intentionally to make cognitive disequilibrium for students in order to facilitate reflective thinking (Davis & Linn, 2000; King, 1991). However, the questions of the teacher are made on the basis of tasks. Thus we must be concerned about the tasks themselves given to students. What are the tasks that may induce cognitive disequilibrium for students? To find an answer for this question, let us return to the aforementioned theories of Dewey and Piaget.

Dewey (1933) suggested five phases of reflective thought, namely (1) suggestions, in which the mind leaps forward to a possible solution, (2) intellectualization of the difficulty or perplexity, (3) hypothesis, (4) reasoning, and (5) testing the hypothesis by action. Among them, what is important in the first phase is experiencing disturbance and perplexity and trying a situation. When faced with a new situation, at first a person undergoes emotional confusion, perplexity and some strange feeling, and such an emotion often leads to reflection.

Piaget (1985) claims that cognitive disequilibrium should induce awareness of puzzling, contradictory, discrepant, or otherwise unassimilable phenomena not previously noticed. There are studies which use paradox as a device for cognitive disequilibrium on the basis of Piaget's theory. According to Rumelhart and Norman (1978) and Kondratieva (2007), a paradox in the broad sense is a suddenly emerged unexpectedness, a statement or situation that looks unbelievable, wrong or contradictory. Since a paradox provides a dis-equilibration, it makes the subject realize the need to re-equilibrate.

In summary, the problem which causes cognitive disequilibrium is inconsistent with the primary thinking and, and it should be able to make students feel perplexity, strangeness or difficulty.

2.4 *The main stages of a process of reflective thinking*

Now we will analyze how and through what process mathematical tasks that can create disequilibrium for students enable reflective thinking and expand students' thoughts and recognition.

Gagatsis and Patronis (1990) investigated how geometrical models can be used by students and teacher in a process of reflective thinking. Although their model applied a process of reflective thinking to a specific geometric task, their model itself is independent from a specific content. So, we will apply this model to our polar coordinates tasks.

Table 4 shows four main stages of a process of reflective thinking. According to this process, reflecting occurs twice; in stages 1 and 3. The difference in the two stages is the level of reflective thinking. While the reflecting in Stage1 is done on previously existing intuitive structures for topical understanding, the one in Stage 3 realizes a higher awareness by using discovery, understanding and one's own mental processes achieved by the reflecting in stage 1 as objects of reflection. In the next section, the process of nurturing reflective thinking will be explained through four polar coordinates tasks using the main stages of a process of reflective thinking in Table 4.

Table 4
The main stages of a process of reflective thinking

Stage	Description
Stage 0 "local" success or failure	Initial thoughts, primary intuitions and conceptions on a subject matter or a problem; starting the work with "local" success or failure; unclear and "fragmentary" mental images; making observations "at random".
Stage 1 Reflecting	**Reflecting on the subject and trying to understand**, i.e. to organize the new experience into previously existing intuitive structures; classifying observations, analyzing wholes into parts, reflecting on them, recalling other similar examples, finding counterexamples, questioning former beliefs and conceptions.
Stage 2 Discovery and (partial) understanding	Discovery and (partial) understanding; finding and/or justifying a rule; finding an explanation for some error; recombining parts of a decomposed whole into new whole interpretation and (partial) reorganization of the new facts according to previous structures; "completion" of mental images and deriving a plan of the solution or of the proof; intuitive feeling of certainty for the success of the plan.
Stage 3 Introspecting	**Introspection; trying to see "what is all about"** i.e. **reflecting** on the process of the solution, the logic of the proof and one's own mental structures and processes; checking or testing one's own results (or conclusions) into other problems or fields; examining analogies and setting up new questions; analyzing the whole situation again, but at a higher level; questioning again; questioning "questioning"; epistemological dispute.
Stage 4 Full awareness	Full awareness; understanding the underlying logic; illumination of the whole subject; becoming aware of one's own mental structures and processes; widening the old structures, transforming or fully rejecting them (and constructing new ones); radical reorganization of ideas, possibly on new foundations; making sound generalizations and extensions; constructing and formulating new theories.

Adapted from Gagatsis and Patronis, 1990, p.33

3 Mathematical Tasks

3.1 *System of equations in polar coordinates*

In rectangular coordinates, the points of intersection are usually found algebraically by solving simultaneous equation algebraically. In the same way, one would expect to find the points of intersection of graphs in polar coordinates algebraically. However, not all points of intersection of Polar graphs can be obtained algebraically. The analysis of the following Task 1 and Task 2 will show the process of reflective thinking that occurs while the points of intersection of the graphs are found by simultaneous equations in a polar coordinate.

Task 1
Find all points of intersection of the curves $r = \theta, r = \theta + 2\pi$

The simultaneous equations $r = \theta$ and $r = \theta + 2\pi$ have no solution. With a graphic tool, it is clear that the two graphs representing the two equations, drawn in Figures 1 and 2, do intersect!

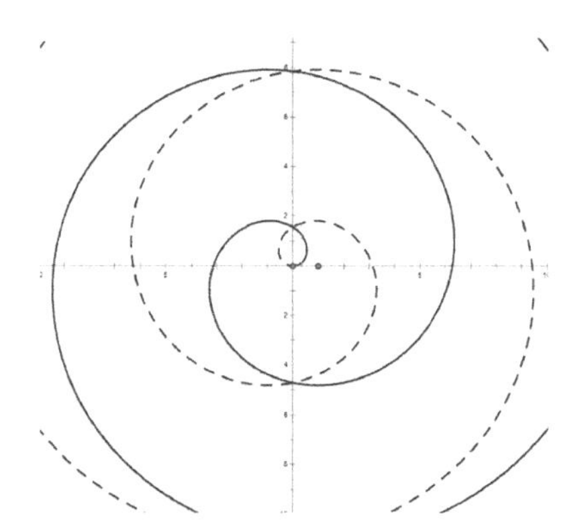

Figure 1. $r=\theta$ (Solid line is when $\theta \geq 0$, and dotted line is when $\theta < 0$.)

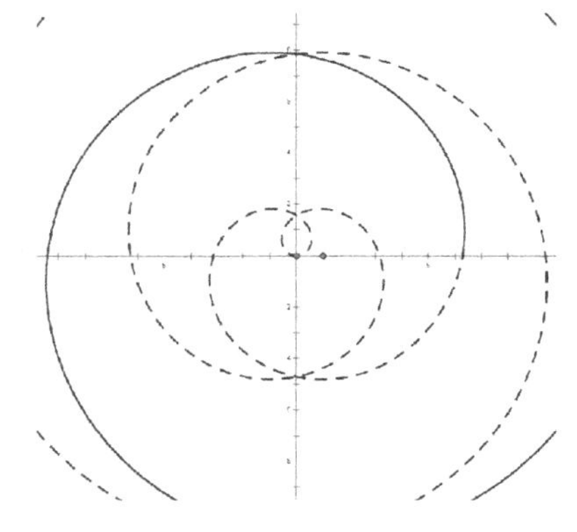

Figure 2. $r=\theta + 2\pi$ (Solid line is when $\theta \geq 0$, and dotted line is when $\theta < 0$.)

Task 2
Find all points of intersection of the curves $r^2 = 4\cos\theta$ and $r = 1 - \cos\theta$.

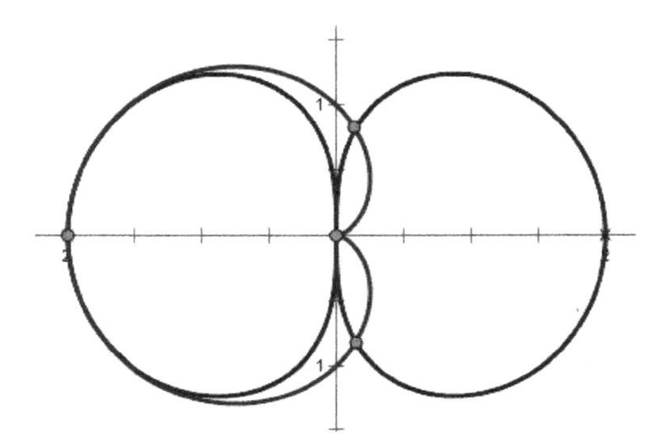

Figure 3. The graphs of $r^2 = 4\cos\theta$ and $r = 1 - cos\theta$

Similarly, we can solve the equations $r^2 = 4\cos\theta$ and $r = 1 - \cos\theta$ and find $r = -2 \pm 2\sqrt{2}$. The value $r = -2 - 2\sqrt{2}$ does not correspond to any point on the curve. The values of θ that meet $r = -2 + 2\sqrt{2}$ are $\pm\frac{4\pi}{9}$. Thus the points of intersection are $(r, \theta)=\left(-2 + 2\sqrt{2}, \pm\frac{4\pi}{9}\right)$. However, as shown in the graph (see Figure 3), $(0, 0)$ and $(2, \pi)$ are also intersection points, though they are not the solution for the equations. The next description shows the process of reflecting about these two tasks.

Process of reflective thinking in Task 1 and Task 2

Stage 0

While solving Task 1 and Task 2, students experience local success or failure when they used the initial conceptions that intersection points on the graph of a rectangular coordinate can be found algebraically with the system of equations. They might not be able to find all the points of intersection of the two graphs in Tasks 1 and 2 algebraically.

Stage 1

First, students reflect on the methods of solving simultaneous equations on a rectangular coordinate that they already know, and compare them with the results of Tasks 1 and 2. As an example of a rectangular coordinate, to get the intersection points of the two graphs of $y = \sqrt{x}$ and $y = x - 2$, we solve the equations, then there are two roots of $x = 1$ or $x = 4$. But $x = 1$ becomes an extraneous root (see Figure 4).

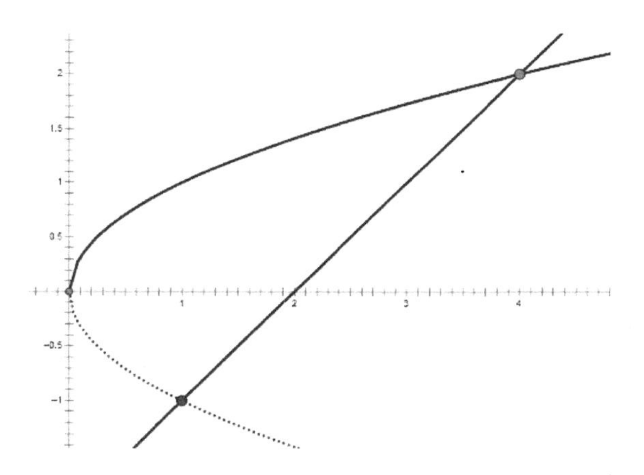

Figure 4. Intersection points in the graphs of $y = \sqrt{x}$ and $y = x - 2$

In a rectangular coordinate system, extraneous solutions may be obtained by solving the equations algebraically. So it is different from the polar coordinates in Task 1 and Task 2 which have insufficient solutions. The problem is different from rectangular coordinates where intersection points of the graphs are found algebraically. Thus one needs to be cognizant that the characteristics of polar coordinates are different from rectangular coordinates. One may begin by analyzing simpler tasks of the type similar to Task 1.

Stage 2

In the equations of $r = \theta$ and $r = \theta + 2\pi$, the difference between θ and $\theta + 2\pi$ is 2π. One should note that a point (r, θ) in the polar coordinates can be expressed in numerous other forms $((-1)^n r, \theta + n\pi)$, where n is an integer.

The graph of a polar equation is the set of all points for which at least one pair of polar coordinates satisfies the equation. To determine a point (r_0, θ_0) lies on the graph of $r = f(\theta)$, it is checked the point may satisfy the equation even while others do not. For $r_0 \neq 0$, if the point (r_0, θ_0) be on the graph, then either $(r_0, \theta_0 + 2n\pi)$ or $(-r_0, \theta_0 + \pi + 2n\pi)$ (for $n \in \mathbb{Z}$) also satisfies the equation. And for $r_0 = 0$, the issue is simply whether the equation $f(\theta) = 0$ has a solution.

Since all the points (r_0, θ_0) that satisfy $r = \theta$ in Task 1 satisfy equation $r = \theta + 2\pi$, the graph $r = \theta$ lies on the graph of an equation $r = \theta + 2\pi$. The converse is also true. Therefore two graphs are identical.

Stage 3

Students should reflect on the logic of the processes in Task 1 and analyze the whole situation again. As the point of $(2, \pi)$ and $(-2, 0)$ marked with a polar coordinate are on curve $r = 1 - \cos\theta$ and $(2, \pi) = (-2, 0)$ is a point on curve $r^2 = 4\cos\theta$, points $(2, \pi)$ and $(-2, 0)$ are both the intersection points of curves $r^2 = 4\cos\theta$ and $r = 1 - \cos\theta$. At this time, $(2, \pi)$ is not satisfied with the equation $r^2 = 4\cos\theta$.

As shown in Task 1, the curves $r^2 = 4\cos\theta$ and $r = 1 - \cos\theta$ also can have different representations. For example, $r = 1 - \cos\theta$ may be expressed by $-r = 1 + \cos\theta$, and $r^2 = 4\cos\theta$ may be expressed by $r^2 = -4\cos\theta$. In order to find the intersection points, it is necessary to solve four different simultaneous equations made by the combination of each of them and in that case $(2, \pi)$ becomes another root. Especially when $r = 0$, $(0, \theta_1) = (0, \theta_2)$ comes into existence values of θ_1, θ_2. which satisfy $0 = 1 - \cos\theta$ and $0 = -4\cos\theta$, so $\left(0, \frac{\pi}{2}\right)$ and (0, 0) are also the intersection points of the two curves.

Stage 4

Reflecting on Tasks 1 and 2, students become aware that they are accustomed to only the problem-solving method of rectangular coordinates; the problems of a polar coordinate need to be considered differently. In polar coordinate system, a point can have multiple representations, which is different from the rectangular coordinate system.

3.2 *Symmetry of polar graph*

Symmetry in the graphs of functions is usually explored in school mathematics. Symmetry is usually approached by geometric transformation such as symmetrical translation, reflection, parallel translation, rotation and congruence. In rectangular coordinates, symmetry of graphs is used in connection with algebraic equation (Ayres & Mendelson, 1999) as shown as follows:

> Consider the graph of an equation $F(x, y) = 0$. Then
> 1) The graph is symmetrical with respect to the y axis if and only if $F(x, y) = 0$ implies $F(-x, y) = 0$
> 2) The graph is symmetrical with respect to the x axis if and only if $F(x, y) = 0$ implies $F(x, -y) = 0$
> 3) The graph is symmetrical with respect to the origin if and only if $F(x, y) = 0$ implies $F(-x, -y) = 0$ (p. 131)

The symmetry of graphs in polar coordinates is used as a convenient method to draw the graph of polar equation. First of all, the method to check for symmetry in polar equation algebraically could be explained (Kwon, Kim, Shin, & Kang, 2011) as follows:

> Consider a polar equation $r = f(\theta)$. Then
> 1) If $f(\theta) = f(-\theta)$ or $(\theta) = -f(\pi - \theta)$, the curve is symmetrical about the polar axis (x-axis).
> 2) If $f(\theta) = f(\pi + \theta)$, the curve is symmetrical about the pole.
> 3) If $f(\theta) = -f(-\theta)$ or $f(\theta) = f(\pi - \theta)$, then the curve is symmetrical about the vertical line $\theta = \pi/2$ (y-axis) (p. 121)

Usually, when sketching the graph of a polar equation, some points which satisfy the polar equation are found and connected smoothly. The steps can be simpler if symmetry is used. For example, in the case of the cardioid $r = 1 + \cos\theta$, since $\cos(-\theta) = \cos\theta$, the curve is symmetrical about the polar axis(x-axis). It suffices to complete the corresponding table of radius r and angle θ from 0 to π and to complete drawing the rest of the graph for θ from 0 to 2π using symmetry (Table 5).

Table 5
Drawing graph in polar coordinates using symmetry

<table>
<tr><td>Step 1. Filling in corresponding table</td><td>Step 2. Drawing graph of $0 \leq \theta \leq \pi$</td><td>Step 3. Completing the rest of the drawing using polar axis symmetry</td></tr>
<tr><td><table><tr><td>θ</td><td>0</td><td>$\frac{\pi}{3}$</td><td>$\frac{\pi}{2}$</td><td>$\frac{2\pi}{3}$</td><td>π</td></tr><tr><td>r</td><td>0</td><td>2</td><td>4</td><td>6</td><td>8</td></tr></table></td><td>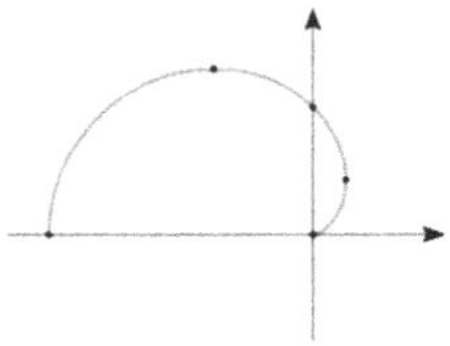</td><td>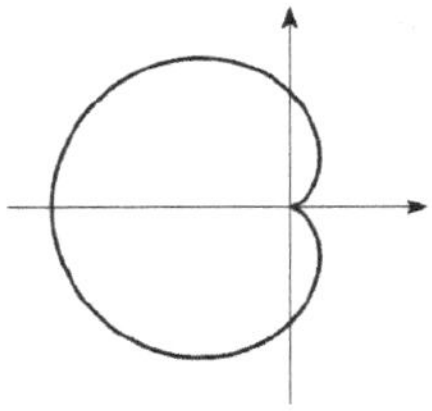</td></tr>
</table>

Task 3 demonstrates the discrepancy in algebraic method to investigate symmetry between the polar and the rectangular coordinate system. Hence the process of learning through reflection becomes useful.

Task 3
Sketch the curve $f(\theta) = \sin\frac{\theta}{2}$.

A student draws the graph of $f(\theta) = \sin\frac{\theta}{2}$ using symmetry. Since $f(-\theta) = \sin\left(-\frac{\theta}{2}\right) = -\sin\left(\frac{\theta}{2}\right) = -f(\theta)$, one would conclude that the graph is symmetrical about the vertical line $\theta = \pi/2$.

Process of reflective thinking in Task 3

Stage 0
Using the same method of sketching a graph of cardioid, a student can complete the graph using symmetry about the vertical line $\theta = \pi/2$ to draw a graph like Figure 5.

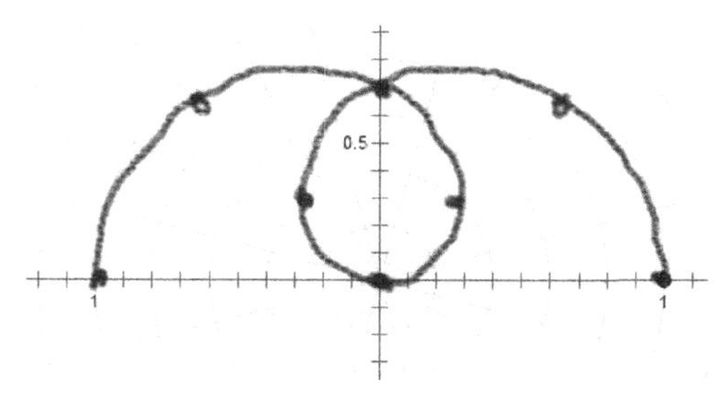

Figure 5. The graph of $f(\theta) = \sin\frac{\theta}{2}$

However, the graph of $f(\theta) = \sin\frac{\theta}{2}$ drawn by graphic software is as shown in Figure 6.

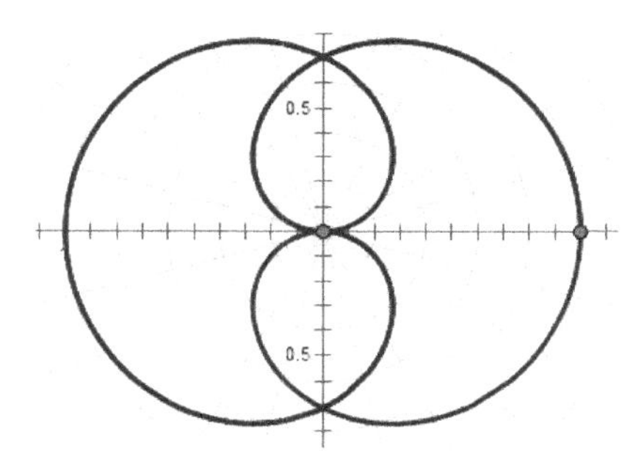

Figure 6. The graph of $f(\theta) = \sin\frac{\theta}{2}$

Only part of the graph was obtained in Figure 6. Here are the observations of the difference between the two graphs.

Stage 1

Based on the fact that only half of the graph was drawn, analysis is made focusing on the period. In $r = 1 + \cos\theta$, a sample given for the drawing of a graph using the symmetry of polar coordinates, the period is 2π. But in $\sin\frac{\theta}{2}$, the period of sine function is 4π instead.

Stage 2

What can be discovered in the observation of Stage1 is the difference in the period of trigonometric function. Therefore, in order to draw a graph $r = 1 + \cos\theta$ using symmetry, the graph should be drawn at least 2π and then make it symmetrical (see Figure 7).

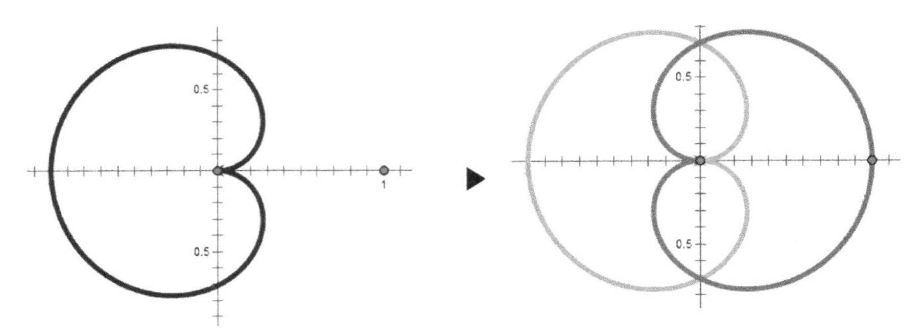

Figure 7. Sketching a graph $f(\theta) = \sin\frac{\theta}{2}$ using symmetry

Stage 3

They look back and compare how symmetry could be used to sketch a graph in a polar coordinate system and a graph in a rectangular coordinate system. As an example of a graph in the rectangular coordinates, consider $y = \cos\left(\frac{x}{2}\right)$, which is symmetrical against the y-axis (see Figure 8). It is sufficient to draw half of the graph on one side of the line of symmetry and then obtain the remainder of the graph by reflection.

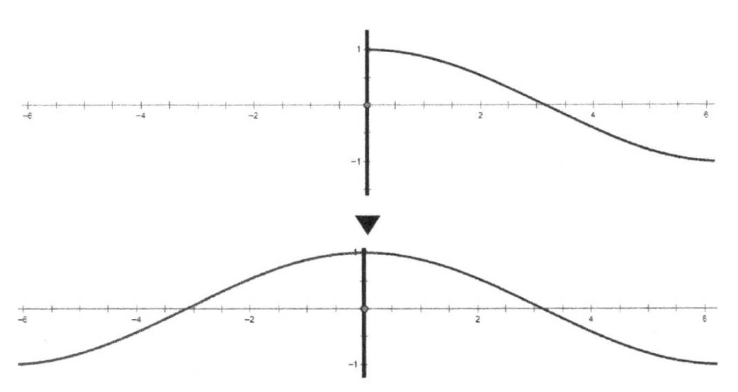

Figure 8. Sketching graph of $y = \cos\left(\frac{x}{2}\right)$ in rectangular coordinates using symmetry

When drawing the graph of a trigonometric function $y = \cos\left(\frac{x}{2}\right)$ in a rectangular coordinate using symmetry, there is no problem in completing the rough figure of the graph even if the graph was not drawn up to 2π, half of the period. To find an explanation for such a difference, students compare the methods used for drawing graphs in the two coordinates.

Let us translate the standard procedure of plotting a graph in rectangular coordinates by plotting few points and joining them up by a smooth curve to polar coordinates. Figure 9 shows the process of drawing a graph of $f(\theta) = \sin\frac{\theta}{2}$. First, make a correspondence table of radius r for several specific angles. The reference line corresponding to the straight line vertical to x- axis in rectangular coordinates is a straight line of $\theta = \theta_0$. Plot a point where it is $\frac{1}{2}$ away from the origin in the positive direction on the straight line of $\theta = \frac{\pi}{3}$. Connect the points counter clockwise to get the graph up to 2π (Figure 9).

θ	0	$\frac{\pi}{3}$	$\frac{\pi}{2}$	π	$\frac{3\pi}{2}$	$\frac{4\pi}{3}$	2π
r	0	$\frac{1}{2}$	$\frac{\sqrt{2}}{2}$	1	$\frac{\sqrt{2}}{2}$	$\frac{1}{2}$	0

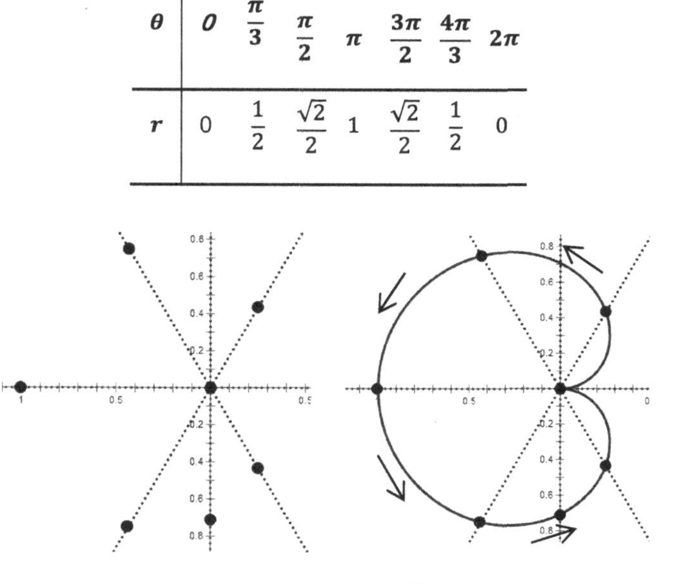

Figure 9. Drawing graph of $f(\theta) = sin\frac{\theta}{2}$ in polar coordinates (1)

After drawing the graph up to 2π, the reference line turns one round on the plane. In rectangular coordinates, it may be thought that the graph has been drawn sufficiently. However, as shown in Figure 10, the reference line of $\theta = \frac{7\pi}{3}$ has an opposite value of $r = -\frac{1}{2}$ though it is the same reference line as $\theta = \frac{\pi}{3}$.

θ	...	$\frac{7\pi}{3}$	$\frac{5\pi}{2}$	...
r	...	$-\frac{1}{2}$	$-\frac{\sqrt{2}}{2}$	...

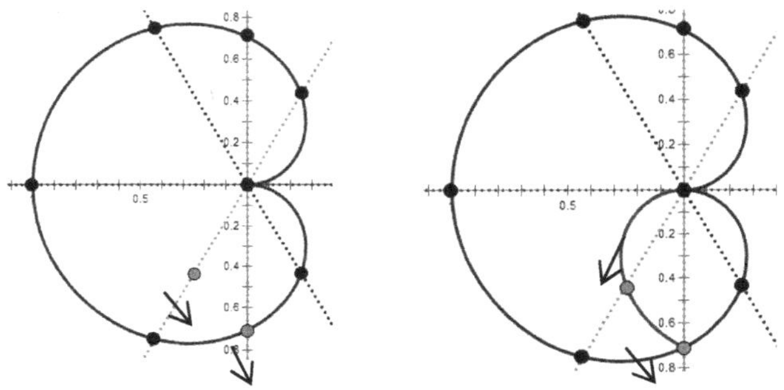

Figure 10. Drawing graph of $f(\theta) = \sin\frac{\theta}{2}$ in polar coordinates (2)

In plotting graphs on both the rectangular and polar coordinate systems, the standard procedure of plotting few points and connecting them up by a smooth curve is generally used. However, in a rectangular coordinate the value of x becomes the reference for drawing the graph as the corresponding y value is found by a specific x value, while in a polar coordinate the value of θ becomes the reference as the corresponding value of r is found by the value of specific θ.

Stage 4

Reflecting the process of drawing the graph of polar coordinates in comparison with that of rectangular coordinates, Students become aware that the difference between the two coordinates comes from the difference in the characteristics of x coordinate and θ coordinate. In rectangular coordinates, any two points with different value of x coordinate become different points (regardless of the value of y

coordinate). However, in polar coordinates, they could be represent the same point even if the value of θ coordinate is different. The points (r, θ) and $(r,\ \theta + 2\pi)$ represent the same point on the plane, although the values of θ and θ+2π are algebraically different.

Unlike graphs in rectangular coordinates, Task 4 demonstrates that there could be discrepancy between the "symmetry" obtained from algebraic equations and the actual symmetries on a graph represented in polar coordinates. This task illustrates how students can be engaged with a process of deeper understanding of the characteristics of polar coordinates by reflective thinking about such difference.

Task 4

Check symmetry of polar graph $f(\theta) = \sin\frac{\theta}{4}$.

To check the symmetry algebraically, since $\mathrm{f}(-\theta) = \sin\left(\frac{-\theta}{4}\right) = -\sin\left(\frac{\theta}{4}\right) = -\mathrm{f}(\theta)$, the polar graph is symmetric about $\theta = \frac{\pi}{2}$ axis(y-axis). It does not satisfy the equation checking the symmetry by polar axis because $-\mathrm{f}(\pi - \theta) = -\sin\left(\frac{\pi-\theta}{4}\right) \neq \sin\left(\frac{\theta}{4}\right) = f(\theta)$ and $f(-\theta) = \sin\left(-\frac{\theta}{4}\right) = -\sin\left(\frac{\theta}{4}\right) \neq \sin\frac{\theta}{4} = f(\theta)$. Similarly $f(\pi + \theta) = \sin\left(\frac{\pi+\theta}{4}\right) \neq \sin\left(\frac{\theta}{4}\right) = f(\theta)$, and it does not satisfy the equation checking the symmetry by the pole (origin).

Process of reflective thinking in Task 4

Stage 0

When students check the symmetry algebraically, it was possible to confirm only the symmetry of $\theta = \frac{\pi}{2}$ axis (y-axis). On the other hand, if they check graphically as in Figure 11, this graph has lines of symmetry about the polar axis, the line $\theta = \frac{\pi}{2}$ and pole symmetry.

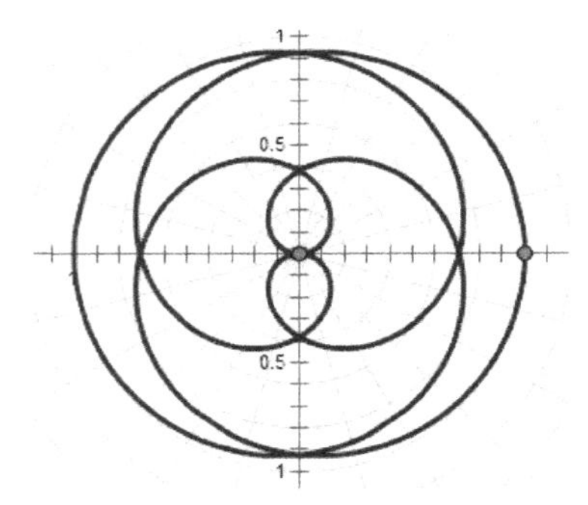

Figure 11. The graph of $g(\theta) = \sin\frac{\theta}{4}$

Stage 1

It can be checked experimentally with the example of another graph. In the case of $f(\theta) = \cos\theta$, it satisfies polar axis symmetric check because $f(-\theta) = \cos(-\theta) = \cos\theta = f(\theta)$, but it cannot be checked for other symmetry. The result of the check by graph is same.

For another example the graph of $f(\theta) = \sin\frac{\theta}{2}$, it can be checked the symmetric about the vertical line $\theta = \pi/2$. But the equation for the symmetry to polar axis or pole is not completed. So algebraic check and graphic expression do not correspond (Figure 12).

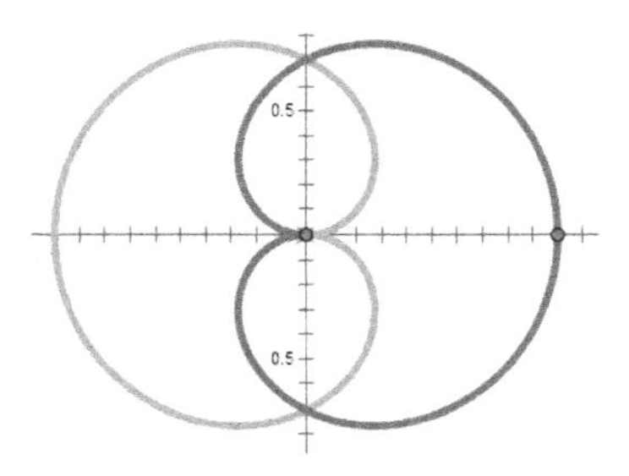

Figure 12. The graph of $\boldsymbol{f(\theta) = \sin\frac{\theta}{2}}$

If we draw several graphs in addition and observe and classify them, we will be able to understand that, unlike a rectangular coordinate, in polar equation it is not always true to say that the graph is not symmetrical because its symmetry is not confirmed algebraically.

Stage 2

One observes that a function f(θ) satisfying $f(-\theta) = -f(\theta)$ or $f(\theta) = f(\pi - \theta)$ in polar coordinates is only a sufficient condition for symmetry about the vertical line $\theta = \pi/2$. In rectangular coordinates, algebraic symmetry is equivalent to graphical symmetry.

Stage 3

Going further from the facts found and understood in stage 2, students can have more questions about them. Why such difference occurs? Is it not possible for being symmetrical in the graph and checking symmetry algebraically to be equivalent both in a polar coordinate and a rectangular coordinate?

Now go back to the point where problems happened in Task 4. As $-f(\pi - \theta) = -\sin\left(\frac{\pi-\theta}{4}\right) \neq \sin\left(\frac{\theta}{4}\right) = f(\theta)$ in general, it is not possible to check the symmetry algebraically. A graph in polar coordinates can be expressed by various equations. In the case of the graph of the equation $r = -\sin\left(\frac{\pi-\theta}{4}\right)$, it can be expressed as $(-1)^5 r = -\sin\left(\frac{\pi-(\theta+5\pi)}{4}\right)$, or $r = \sin\frac{\theta}{4}$. It is same with the symmetry by extreme point. In the symmetry by pole, it is $f(\pi + \theta) = \sin\left(\frac{\pi+\theta}{4}\right) \neq \sin\left(\frac{\theta}{4}\right) = f(\theta)$. Though the equation is not complete algebraically, in terms of graphic expression, the graph of the equation of $r = \sin\left(\frac{\pi+\theta}{4}\right)$ is the same as the output of the graph made by $(-1)^3 r = \sin\left(\frac{\pi+(3\pi+\theta)}{4}\right)$, or $r = \sin\frac{\theta}{4}$.

On the basis of the results found by reflective thinking, it is considered more about the method used to check the symmetry by polar axis algebraically. Various textbooks of calculus present:

> If polar equation is unchanged when θ is replaced by $-\theta$ or $\pi - \theta$, the curve is symmetric about the polar axis (x-axis) (Stewart, 2006, p. A9).

> If the same polar equations obtained by replacing θ by $-\theta$ or by replacing θ by $\pi - \theta$, and r by $-r$, then the curve is symmetric about the x-axis (Fraleigh, 1990, p. 581).

If $f(\theta) = f(-\theta)$ or $(\theta) = -f(\pi - \theta)$, the curve is symmetric about the x-axis (Kwon, Kim, Shin, & Kang, 2011, p. 121).

Rather than restricting the change of a polar equation when θ is changed into $-\theta$ or $\pi - \theta$ in a polar equation to check for invariance to check for symmetry of the graph, the condition could be changed to make the graph of a polar equation invariant through other substitution. Or there will be the same equation in the polar equations showing the graph of $r = f(\theta)$ and the polar equations showing the graph of $r = f(-\theta)$ or $r = -f(\pi - \theta)$.

Stage 4

Summarizing mathematical tasks presented in the paper, the fundamental difference between rectangular coordinates and polar coordinates can be generalized. In rectangular coordinates, a point on the plane and an expression for it make one-to-one correspondence. As a result, generally an algebraic equation and the graph also make one-to-one correspondence. However, in a polar coordinate, there are numerous expressions used to refer to a point. As a result, it becomes possible to have 'many different' polar equation expressions to show a polar graph. It is clearly recognized that all other polar equations must be considered when dealing with polar graphs.

4 Conclusion

This study intended to show that the tasks provoking cognitive disequilibrium provide students with the opportunity to complement their incomplete understanding through reflective thinking. To be able to find appropriate tasks, a teacher must have an in-depth understanding of the nature and characteristic of the mathematical concepts that s/he is supposed to teach. In addition, the role of the teacher in actual class is very important. Presenting such asks is the beginning for the disequilibrium which prompts students to have reflective thinking. In this case, the teacher must provide proper scaffolding through questioning to improve recognition of disequilibrium. The teacher will have to guide students to embody the attitude of mathematical thinking checking strategies and methods of mathematical thinking for themselves, and furthermore find the joy of mathematical discovery through experience of problem-solving independently.

References

Ayres, F., & Mendelson, E. (1999). *Schaum's outline of calculus*. New York: McGraw-Hill.

Boyer. C.B. (1949). Newton as an originator of polar coordinates. *The American mathematical monthly, 56*(2), 73-78.

Davis, E.A., & Linn, M. (2000). Scaffolding students' knowledge integration: Prompts for reflection in KIE. *International Journal of Science Education, 22*(8), 819-837.

Dewey, J. (1933). *How we think: a restatement of the relation of reflective thinking to the educative process*. Chicago: Henry Regnery Company.

Fraleigh, J.B. (1990). *Calculus with analytic geometry (3rd ed.)*. Reading, Mass.: Addison-Wesley Publication Co.

Freudenthal, H. (1978). *Weeding and sowing: Preface to a science of mathematical education*. D. Reidel Publishing Company.

Freudenthal, H. (2002). *Revisiting mathematics education China lectures*. NY: Kluwer Academic Publishers.

Gagatsis. A., & Patronis, T. (1990). Using geometrical models in a process of reflective thinking in learning and teaching mathematics. *Educational Studies in Mathematics, 21*, 29-54.

Ge, X., & Land, S.M. (2003). Scaffolding students' problem-solving processes in an ill-structured task using question prompts and peer interactions. *Educational Technology Research and Development, 51*(1), 21-38.

King, A. (1991). Effects of training in strategic questioning on children's problem-solving performance. *Journal of Educational Psychology, 83*(3), 307-317.

Kondratieva, M. (2007). Understanding mathematics through resolution of paradoxes. *Mediterranean Journal for Research in Mathematics Education, 6*, 127-138.

Kwon, O.N., Kim, S.R., Shin, D.Y., & Kang, H.J. (2011). *Fundamentals of advanced mathematics*. Seoul: Seoul textbook.

Ministry of Education, Science and Technology. (2011). *Mathematics Curriculum* [Notice no: 2011-361, separate volume 8]. Republic of Korea: Ministry of Education, Science and Technology.

Piaget, J. (1976). *The grasp of consciousness*. Cambridge, MA: Harvard University Press.

Piaget, J. (1977). *The development of thought: Equilibration of cognitive structures*. New York: Viking.

Piaget, J. (1985). *The equilibration of cognitive structures*. Translated by Brown, T. *et al.*, The University of Chicago Press.

Polya, G. (1973). *How to solve it*. Princeton, NJ: Princeton University Press.

Rumelhart, D.E., & Norman, D.A. (1978). Accretion, tuning and restructuring: three modes of learning. In, J.W. Cotton & R.L. Klatzky (Eds.), *Semantic factors of cognition* (pp. 37-52). Hillsdale, N.J.: Lawrence Erlbaum Associates.

Skemp, R. (1979). *Intelligence, learning, and action*. New York: Wiley.

Stewart, J. (2006). *Calculus: Concepts and contexts (3rd ed)*. CA: Thomson Brooks/Cole.

Chapter 12

Using Codes to Facilitate Metacognition in Mathematics

WONG Oon Hua

This chapter looks at an alternative method of providing feedback to primary four pupils in a mathematics class in Singapore. Traditionally, teachers often use ticks and crosses to grade mathematics assignments. A tick means that the solution is correct while a cross means that the solution is incorrect. This process of grading does not give a pupil any information about possible errors or gaps in his or her solution. In this chapter, I will examine an alternative method of giving pupils feedback about their written assignments. This method involves the use of codes to indicate types of errors a pupil may have made in his or her written work. In so doing, pupils are given an opportunity to engage in thinking about their errors. I will also discuss some findings from a small scale study on the use of codes for feedback to pupils about their written work in mathematics.

1 Introduction

In many classrooms in Singapore it is a common practice in language instruction, particularly English Language, for teachers to provide feedback to learners about their written composition exercises using codes. An example of a set of codes commonly used is shown in Figure 1. Teachers often draw the attention of pupils to the errors they have made in their written compositions by underlining the error and indicating with a code the type of error made. As shown in Figure 2, the pupil has made numerous types of errors in his or her written composition and the teacher has marked out the errors and labeled them accordingly. The teacher has drawn the attention of the pupil to the kind of error, for example 'S' or 'A', etc., made but did not correct the error for the pupil. This type of feedback is significant as it allows the pupil to correct his or her own errors with sufficient guidance. It also allows the teacher to note the main types of errors pupils are making and tailor instruction to address gaps in their learning.

Can the same idea be used for grading mathematics written assignments? Traditionally, teachers often use ticks and crosses to grade mathematics assignments. A tick means that the solution is correct while a cross means that the solution is incorrect. This process of grading does not give a pupil any information about possible errors or gaps in his or her solution. What does the literature say about the use of codes to facilitate metacognition? Is there anything else that we can learn from the literature regarding the traditional manner of marking using ticks and crosses as a form of feedback? In the next section we explore some of these concerns.

English Language Marking Codes

S – Spelling
P – Punctuation (full stop, comma, question mark, apostrophe, etc.)
PP – Preposition (on, in, under, over, beside, behind, etc.)
SV – Subject verb agreement (The boy eats. The boys were going home.)
W – Choice of word
A – Article (a, an, the)
T – Past tense (eat – ate, teach – taught, speak – spoke, has – had)
PT – Present tense
N – Number (two apple – two apples, four childrens – four children)
^ – Missing word
? – Unclear

Figure 1. An example of a set of codes for grading English Language composition

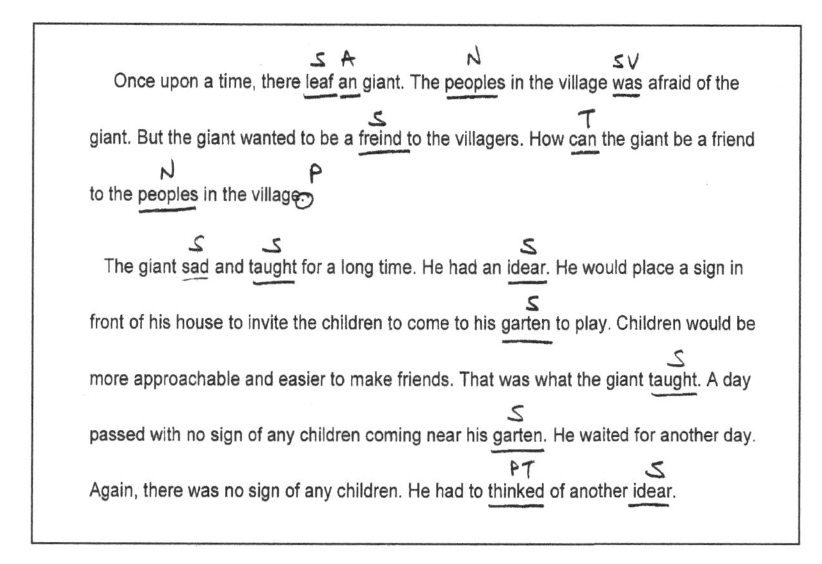

Once upon a time, there leaf an giant. The peoples in the village was afraid of the giant. But the giant wanted to be a freind to the villagers. How can the giant be a friend to the peoples in the village

The giant sad and taught for a long time. He had an idear. He would place a sign in front of his house to invite the children to come to his garten to play. Children would be more approachable and easier to make friends. That was what the giant taught. A day passed with no sign of any children coming near his garten. He waited for another day. Again, there was no sign of any children. He had to thinked of another idear.

Figure 2. Sample of pupil's composition graded using codes

2 Literature Review

The Report of the Primary Education Review and Implementation Committee in Singapore (PERI, 2009, p. 35), recommends "more widespread use of assessment to provide qualitative feedback to students on their learning and development of skills in both academic and non-academic areas, and to suggest how they can make improvements". This certainly signals a need to look beyond ticks and crosses, marks and grades in our feedback to pupils in their regular assignments to help them understand their errors and how to move forward in their learning. In the report one of the ideas suggested is "to use assessment tools such as rubrics to assess and provide pupils with richer feedback on their development in academic and non-academic areas" (PERI, 2009, p. 36). Rubrics may take many forms and it may be claimed that codes serve the same purpose as rubrics in providing pupils with specific guidance regarding their errors and how to proceed to correct them.

Vercauteren (2008) supports the idea of using codes to provide students with feedback that gives specific information about their current achievement, the next step (or goal) and how to reach that goal. For example, the codes indicated in the assignment will provide the specific information about the pupil's current achievement which could be 'W' for working or 'U' for units. The pupil will then be able to look at the next step, to check if there is a computation error for 'W' or wrong use of units for 'U' to reach the goal of mastery of the computation skill or comprehending the appropriate use of units. Callingham (2008) claims that feedback must provide students with two pieces of information; i) affirmation of what they can currently do, and ii) what they need to do next to improve their understanding. It is apparent that both Vercauteren (2008) and Callingham (2008) share the fundamental intent of feedback, namely to provide information to pupils to help them know what their errors are and what they can do to improve their learning. In addition, Brookhart (2008) states that effective feedback contains information that a student is able to understand so that he or she is able to act in a meaningful manner.

Black and Wiliam (1998) further add that "for assessment to function formatively, the results have to be used to adjust teaching and learning" (p. 5). This certainly affirms the dual role of codes, firstly engaging pupils in identifying their errors and correcting them and secondly providing teachers with data on aspects of learning deficiencies their pupils are having, so that they may provide suitable instruction to help their pupils improve their learning. Wiliam (2011), cautions that "feedback functions formatively only if the information fed back to the learner is used by the learner in improving performance" (p. 120). Also, Wiliam (2011) attests that "feedback should cause thinking" and "feedback needs to direct attention to what's next" (pp. 127-128). In using codes, the learner is alerted to the type of error and hence provided with an opportunity to analyse a specific error.

3 A Study on the Use of Codes for Grading Mathematics Assignments

The author carried out a small scale study in which he devised a set of codes, trialled them and subsequently revised them for the grading of written assignments in a primary four mathematics classroom. The following sections give details on the codes devised and implementation of codes for use by pupils.

3.1 *Devising the codes*

Teachers often craft marking schemes to guide them when grading pupils' written assignments or test scripts. The marking scheme spells out clearly how the mark for a question or task would be apportioned between the answer and the method. In so doing, the answer mark takes care of the product while the method mark does the same for the process. The answer is important as we want pupils to be able to arrive at the correct answer at the end of the solution. It is also easy to set up a code for the answer, 'A'. If the final answer is wrong, the teacher will indicate with the code A to signal to the pupil to look at the solution to analyse where the error is.

However, the same cannot be done for the wrong method as an incorrect method may result from several aspects such as the incorrect use of an operation, wrong computation, erroneous transference of data, incorrect use of a heuristic, wrong representation (model) of information, incorrect use of units or missing units, etc. Hence, after much deliberation and consultation with colleagues and scoring guides used by teachers, altogether eight codes were assembled for coding an incorrect method. They were 'U' for an incorrect use of a unit or a missing unit; 'NS' for an incorrect number sentence; 'S' for a final incorrect mathematical statement, 'OP' for incorrect operation used, '>' to indicate a missing step, 'M' for an incorrect, incomplete or missing representation (Model), 'W' for errors arising from calculation or specific strategies and 'T' for transference errors. Figure 3 shows the list of codes devised for coding written mathematics work.

Maths Marking Codes

U – unit (missing, wrong unit) Eg : Ali weighs 30 cm. (30 kg)
NS – Number sentence
S – Statement
OP – Operation (+ , - , x , ÷)
> : missing step
A : wrong answer
M : model drawing Eg : Model drawing needed, incomplete model drawing, wrong model drawing
W : check the working Eg : Specific strategies, calculation errors
T : transfer error Eg : Misread information

Examples

1) 45 + 8 = 125 (A)
2) Rachael has 7 books. Siva has twice as many books as Rachael. How many books do they have?
   7 + 2 = 9 (OP)
   9 + 7 = 16 (NS)
   They have 16 books. (A)
3) Keith and Rick have 30 books. Keith has 10 books. Marcus has 5 more books than Rick. How many books do the 3 boys have?
   30 + 10 + 5 = 45 (NS)
   ➢ (Missing step)
   The 3 boys have 45 books. (A)
4) Ethan has 20 pens. Zee How has 30 pens. How many pens do they have?
   20 + 10 = 50 (T)
   They have 50 altogether. (U)

Figure 3. Set of codes for grading mathematics written work

3.2 *Implementing the codes*

Two groups of students were involved in the study. The first group was the one given feedback about their mathematics written work using codes and we called this group the enhanced feedback group (EFG). The second group was given feedback as per normal practice of the school had been for some time and we called this group the traditional feedback group (TFG). The pupils in the TFG and EFG were randomly assigned and had comparable mean scores based on their end of primary three mathematics examination. The mean score for the TFG was 79.8 and the EFG was 77.7. During the period of the study pupils were doing multiplication and division of whole numbers and solving word problems.

Training pupils to use the codes. Pupils in the EFG were given the set of codes shown in Figure 3 and explained what each of the codes meant. The use of the codes was also illustrated by the teacher using examples of incorrect written work.

Feedback to pupils in the study. During the period of the study pupils were given written work to do on teacher made worksheets. An example of such a worksheet is shown in Figure 4. The worksheets clearly stated the strategies pupils had to use when doing their work. They also allowed the teacher to track the number of times the pupils attempted the corrections. This allowed the teacher to check for effectiveness of the feedback in focusing the pupil on errors made and self-correction. Figure 5 shows the kind of feedback, only ticks and crosses, given to the pupils in the TFG while Figure 6 shows the feedback given using codes to pupils in the EFG. Figure 5 also shows the two attempts the pupil made to correct his error but in vain. Figure 6 also shows the two attempts the pupil made and successfully corrected his error in the second attempt.

Maths Codes – Exercise 3
Name : ______________________ Pr __/___ Date : _________

No of times returned : Well Done! 1 2 3

Strategies : Underline working and answer , check answer ,
simplify the number into groups of tens , use of the equal sign

Example : 48 x 70 (W) – underline the number to work at

= 48 x 7 x 10 (W) – underline the numbers to work at , simplify correctly

= 336 x 10 (W) – underline the numbers to work at , multiply correctly

= 3360 (A) – check the multiplication

1. 32 x 40

Figure 4. A sample worksheet

Maths Codes – Exercise 3
Name : ______________________ Pr __/___ Date : _________

No of times returned : Well Done! 1 2 3

Strategies : Underline working and answer , check answer ,
simplify the number into groups of tens , use of the equal sign

Example : 48 x 70 (W) – underline the number to work at

= 48 x 7 x 10 (W) – underline the numbers to work at , simplify correctly

= 336 x 10 (W) – underline the numbers to work at , multiply correctly

= 3360 (A) – check the multiplication

1. 32 x 40

32 x 40	32 x 40	32 x 40
= 32 x 4 x 40	= 32 x 4 x 10	= 32 x 4 x 40
= 126 x 10	= 138 x 10	= 128 x 4 x 10
= 1260 X	= 1380 X	= 512 x 10 X
		= 5120

2. 190 x 30

Figure 5. Feedback given to pupils in TFG

Maths Codes – Exercise 3
Name : ____________________ Pr __ / __ Date : ________

No of times returned : Well Done! (1) (2) 3

Strategies : Underline working and answer , check answer ,
simplify the number into groups of tens , use of the equal sign

Example : 48 x 70 (W) – underline the number to work at

= 48 x 7 x 10 (W) – underline the numbers to work at , simplify correctly

= 336 x 10 (W) – underline the numbers to work at , multiply correctly

= 3360 (A) – check the multiplication

1. 32 x 40

32 × 40 ✓ | 32 × 40 ✓ | 32 × 40 c
= 32 × 4 × 40 (W) | = 32 × 4 × 10 ✓ | = 32×4×10 c
= 126 × 10 (W) | = 138 × 10 (W) | = 128 × 10 c
= 1260 (A) | = 1380 (A) | = 1280 c

2. 190 x 30

Figure 6. Feedback given to pupils in EFG

3.3 *Data collection and findings*

As part of the study, data was collected from both teachers and students. The data collected from the teachers comprised of the number of corrections pupils made to reach a correct solution in the worksheets and the test scores of the pupils for the four tests during the period of the study. The data collected from the pupils in the EFG comprised a survey about feedback given using codes for their written work.

Findings from the teacher's data. Table 1 shows the percentage of students and the number of corrections made by them to successfully complete the worksheets for the entire duration of the study. From Table 1, it is apparent that 34.8% of the pupils in the EFG were able to correct their mistakes using the codes compared to 23.3% of the pupils in the TFG after one correction. The codes may have helped the pupils in the EFG to better understand their errors and hence more pupils in the EFG compared to the TFG corrected their errors after one round of correction. However, 16.9% of pupils in both groups needed 2 to 3 rounds of corrections. Perhaps some pupils need more guidance in reviewing their misconceptions.

Table 1
Percentage of pupils versus the number of corrections

Group	No of pupils	Number of times pupils made corrections			
		0	1	2	3
TFG	20	59.4%	23.3%	7.8%	9.1%
EFG	21	48.3%	34.8%	6.5%	10.4%

Table 2 shows the mean test scores and standard deviations of the pupils for the four tests. Although it appears from Table 2 that the mean scores of pupils in the EFG were higher than those in the TFG for Tests 1, 2 and 3, there were no statistical differences between the means of the two groups at the 95% confidence level. Tests 1, 2 and 3 were based on the sub-topics multiplying a number up to 3-digit by tens, multiplying a 2-digit number by a 2-digit number and dividing a 3-digit number by a 1-digit number respectively. Test 4 was on solving up to 3-step word problems involving the 4 operations. Both the TFG and EFG scored an average of less than half of the maximum possible marks for Test 4. Again there was no statistical difference between the means of the two groups.

Table 2
Mean and standard deviation of the tests

Group	Mean score (standard deviation)			
	Test 1	Test 2	Test 3	Test 4
TFG (n = 20)	9.15 (1.60)	9.20 (0.86)	9.30 (1.22)	4.83 (1.70)
EFG (n = 21)	9.86 (0.36)	9.38 (1.12)	9.43 (0.75)	4.45 (1.52)

N.B. Maximum possible score for each test is 10.

Findings from the pupils' data. Table 3 shows the data from the pupils' survey that was administered to the EFG. Pupils were asked to put a tick in the box that best describes their experiences in using the codes to correct their errors for written mathematics work. From Table 3, it is apparent that at least 80% of the pupils strongly agree or agree with all the statements in the survey. All the pupils strongly agree or agree that they find the mathematics codes helpful, enjoy learning mathematics using the codes and would want to continue using the codes. 95.2% strongly agree or agree that they are able to correct the errors using the codes, easy to use the codes and will keep trying when they have difficulty in mathematics. Thus the data in Table 3 show that the EFG pupils found the use of codes as a form of feedback for correcting their errors in written mathematics indeed helpful and meaningful.

Table 3
Survey data from EFG about use of maths codes for feedback on written work

No	Item	SA	A	D	SD
1	I find the use of the maths codes to be helpful.	9 (42.9%)	12 (57.1%)	0 (0%)	0 (0 %)
2	I am able to correct my errors using the maths codes.	12 (57.1%)	8 (38.1%)	1 (4.8%)	0 (0%)
3	The maths codes help me to reflect on my learning.	7 (33.3%)	10 (47.6%)	4 (19.0%)	0 (0%)
4	I enjoy learning maths using the maths codes.	12 (57.1%)	9 (42.9%)	0 (0%)	0 (0%)
5	I am more confident in maths.	11 (52.4%)	6 (28.6%)	4 (19.0%)	0 (0%)
6	I want to continue using the maths codes.	13 (61.9%)	8 (38.1%)	0 (0%)	0 (0%)
7	I am more aware of the process in doing maths.	7 (35.0%)	11 (55.0%)	2 (10.0%)	0 (0%)
8	I know what I need to look out for in doing maths.	14 (66.7%)	5 (23.8%)	2 (9.5%)	0 (0%)
9	It is easy to use the maths codes.	15 (71.4%)	5 (23.8%)	1 (4.8%)	0 (0%)
10	I will keep trying when I have difficulty in maths.	10 (47.6%)	10 (47.6%)	1 (4.8%)	0 (0%)

Legend: SA-Strongly Agree, A-Agree, D-Disagree, SD-Strongly Disagree

4 Some Considerations for the Use of Codes

If pupils have been using codes in analysing their errors in composition, then the transition to use codes in mathematics will be less problematic. Otherwise more effort would need to be expanded in getting pupils to acquaint with the concept and process. Firstly, the teacher will have to make an important decision on the types and number of codes to use with the pupils at the start so that the pupils are comfortable and confident in recognising and understanding the codes. Next, a practice session will be helpful to familiarise the pupils with the codes and a simulated exercise on how to use them. Once the familiarization phase is over, given the codes the pupils will be able to analyse their errors, think about them and ultimately correct them. For example, when adding 2 decimals, 0.6 + 0.05 the error code 'W' for working, signals a pupil to look at the place values or the renaming of the decimals or checking that the addends have the same number of decimal places. Furthermore, the codes provide a common language and platform for pupils to discuss their errors.

The teacher may organise the pupils to be in Group W for those with working errors or Group U for those with units errors or a mixed group for pupils with any error so that they may help one another. The teacher can also keep track of the types of errors pupils make and plan remediation sessions accordingly. For the pupils, by keeping track of the frequency of the types of errors they make, they will be able to monitor their progress in learning. This will motivate them to reduce the types of errors they make. Notably, most of the pupils in the EFG remarked that codes *make it easier to understand our mistakes* and codes *inform us where we went wrong so that we can correct our mistakes easily*. This may be said to sum up the essence of metacognition, i.e., to help pupils think about their thinking (Ministry of Education, 2006).

5 Conclusions

This chapter has shown how codes may be used to provide pupils with feedback about their errors in written mathematics work. The idea was borrowed from the coding of errors in English Language compositions by some Language teachers. It may be said that the process of using codes is like de-coding a message which guides you but does not give you the correct 'answer'. This step is critical as the pupil must use the guide and figure out the correct answer by thinking through the solution process. Therefore, it provides an opportunity for pupils to engage in thinking about their errors and correcting them by analyzing the code indicated by the teacher. This is synonymous to "teaching the child to fish" and not "giving him or her the fish". As the famous Chinese saying goes: Give me a fish and I eat for a day, teach me how to fish and I will eat for life! Hence it is important that we empower our pupils to be self-directed learners and confident pupils.

Acknowledgements

The author is grateful to Associate Prof Wong Khoon Yoong for his constructive feedback on the first draft of the chapter. He is also indebted to Prof Berinderjeet Kaur for all the help she rendered in writing the chapter.

References

Black, P., & Wiliam, D. (1998). *Inside the black box: Raising standards through classroom assessment.* London: King's College London School of Education.

Brookhart, S.M. (2008). *How to give effective feedback to your students.* Alexandria, VA: Association for Supervision and Curriculum Development.

Callingham, R. (2008). Dialogue and feedback: Assessment in the primary mathematics classroom. *Australian Primary Mathematics Classroom, 13*(3), 18-22.

Ministry of Education. (2006). *Mathematics syllabus – Primary.* Singapore: Author.

PERI (2009, March). *Report of the primary education review and implementation committee*. Singapore: Ministry of Education.

Vercauteren, D. (2008). *Teacher feedback to primary school students: Do they get the message?* La Vergne, TN: Lightning Source Inc.

Wiliam, D. (2011). *Embedded formative assessment*. Bloomington, IN: Solution Tree Press.

Chapter 13

Mathematics Lessons Stimulating Reflective Learning: Japanese Perspective

Keiko HINO

Aiming at fostering abilities for independent doers of mathematics, we have been emphasizing the process of solving problems in mathematics lessons in Japan. This chapter focuses on Japanese problem solving approach to mathematics teaching and examines how we try to stimulate children's reflective thinking and learning in the classroom. Some key Japanese words that describe perspectives of stimulating reflection through structured problem solving are introduced. They are "Kizuki" (noticing), "Toi" (question), "Neriage" (kneading up) and "Yosa" (merit of mathematical ideas). In the latter part of the chapter, teachers' trials of stimulating reflection in mathematics lessons are described. These trials show that there are a variety of roads to stimulating children's reflections to teach mathematics and that purposefully designing, implementing and evaluating such lessons is an important vehicle for teachers' professional development.

1 Introduction

In Japan, we have national curriculum standards, called the Course of Study (CS), that is prescribed by Ministry of Education, Culture, Sports, Science and Technology (MEXT). After World War II, the CS has been revised approximately every 10 years. The history of the CS shows that although the emphasis in the CS varies in each revision, we have been emphasizing the abilities of thinking mathematically and utilizing mathematics in real world, and in communication. The process of mathematical problem solving provides a powerful method by which to foster these abilities in the teaching of mathematics in the classroom. Many studies have been conducted on mathematical problem solving and we continue to investigate the classroom conditions in which children are vitally engaged in the activity of solving problems (Hino, 2007).

This chapter focuses on the problem solving approach to mathematics teaching in Japan and, in particular, examines how we try to stimulate children's reflective thinking and learning by solving problems in the classroom. In doing so, the author first describes the Japanese problem solving approach with special attention to the roles the teacher played in the development of the lesson. Next, some keywords are introduced that describe the approaches for stimulating children's reflection through structured problem solving. An example of textbook problem is used to show how a lesson is organized from these perspectives. Two practice-based studies that were conducted by groups of teachers in an in-service education program are then described. The intent of the description is to show that there are a variety of possible ways to stimulate children's reflections in the classroom, and that purposefully designing, implementing and evaluating such lessons is an important vehicle for teachers' professional development.

2 Teaching Mathematics in the Classroom: Problem Solving Approach

2.1 *Japanese lesson pattern*

Needless to say, mathematics is not taught solely by presenting lectures. It is almost impossible to teach mathematics without letting children themselves solve mathematical problems. In fact, according to the results of TIMSS 1999 videotape study (Hiebert *et al.*, 2003), one of the general features in the seven participating countries was the teaching of mathematics through solving problems. The results show that at least 80% of lesson time, on average, was devoted to solving mathematics problems. However, a closer look of the results reveals differences among countries in the relative emphasis they placed on different values in organizing lessons around solving problems.

It was in the context of the TIMSS 1995 videotape study that Japanese lesson pattern was described together with the German and U.S. lesson pattern (Stigler & Hiebert, 1999). The term, lesson pattern, refers to the recurring features of mathematics lesson that define different sections of a lesson and the method by which the sections are sequenced. The Japanese lesson pattern was described as a sequence of five activities (or stages):

- Reviewing the previous lesson,
- Presenting the problem for the day,
- Students working individually or in groups,
- Discussing solution methods, and
- Highlighting and summarizing the major points (Stigler & Hiebert, 1999, pp. 79-80).

As the sequence of five activities indicates, the Japanese lesson pattern provides children with the time to solve the problem on their own in an early section of the lesson. Then the children present their solution methods and discuss them with their classmate and with the teacher in a whole-class activity. Based on the proposed solutions, new mathematical content is introduced to children, which is the objective of the lesson. Stigler and Hiebert called this Japanese way of teaching mathematics in the classroom as *structured problem solving*.

Development of the lesson based on these five activities is described in detail by Stigler and Hiebert (1999, pp. 36-41). Associated with descriptions of Japanese lesson pattern, Shimizu (2009a) points out the metaphor of a story or a drama that underlies an excellent lesson. Shimizu noted that, "a lesson needs to have a highlight or climax based on the active role of students guided by the teacher in a coherent way" (p. 95). Accordingly, the Japanese lesson pattern offers a useful perspective from which to investigate the features of coherence in the classroom (Shimizu, 2009a; Hino, 2009; Sekiguchi, 2012). So, how is it possible to implement a lesson effectively as structured problem solving?

In the next sub-section, I describe the five activities with special attention to the roles of the teacher in developing the lesson.

2.2 *Critical roles played by teacher in the structured problem solving*

Given that an actual lesson last for 45 minutes, it is not an easy task to implement the structured problem solving in a smooth manner. The role of the teacher is crucial. Here, specific roles played by teacher are described in relation to each of the five activities that characterize Japanese mathematics lessons.

Reviewing the previous lesson. The lesson usually begins with review of the previous lesson. The teacher asks questions such as "What did you learn in the previous lesson?" or "How did you investigate the problem yesterday?" By recalling the topic, content or method of investigation that they studied in the previous lesson, the teacher arouses children's attention to the connection between that lesson and today's lesson. The teacher often summarizes what was learned by writing important knowledge on the blackboard. For the purpose of eliciting the connection with the previous lesson, the teacher sometimes uses children's comments or questions that were heard during the previous lesson or recorded in their notebooks.

Presenting the problem for the day. In posing the problem for the day, the teacher presents the problem which is to be dealt with in depth that day. The teacher may use different teaching materials to elicit children's interest. She/he may again ask questions to draw children's attention to the connections between the previous lessons and today's lesson, for example, "Which part do you think is the same and which is different from the problems in the previous lesson?"

Another important role of the teacher is to enhance children's understanding of the problem. Here, it is important to ask for children's perspectives or conjectures both on the answer and on the solution method. Therefore, teacher can ask questions such as, "What do you think the answer is?" or "Do you have any idea of how to approach this problem?" This initial thinking often produces a variety of conjectures from children. By listening to the variety of answers children have the opportunity to compare their thinking with their classmates, which contributes to enhancing their understanding of the problem.

Students working individually or in groups. In this stage children tackle the problem in their own way. Since they only have the knowledge and skills they have acquired thus far, this activity is a genuine opportunity to use their knowledge to deliver a plausible answer. Sugiyama and Ito (1990) say that this activity enables children to experience the problem solving process and, especially, the confidence and joy of being able to find a solution to the problem by drawing on their own knowledge. However, the teacher should know that even during the time of this individual problem solving, children need to be assisted by the teacher in order to spend their time productively.

The teacher, therefore, plays an important role while she/he walks around the desks looking at and listening to children's ideas and difficulties. According to his/her observation of individual children, the teacher offers hints or prompts, encourages children to think of various ways to solve the problem, or provides extended tasks for those who have finished the task. The teacher needs to be flexible in managing the time. When many children are found to be having difficulty in approaching the task, the teacher may stop the activity and take time to check their understanding or to ask some children to share their ideas about how to approach the task with their classmates. Another role of the teacher is to analyze children's thinking to solve the problem. The purpose here is to think about the order of presentations and about the ways of organizing discussion that will take place later in the class. These roles are summarized by Shimizu (1999) as "assessing students' problem-solving progress" and "purposeful scanning of their solutions."

It is often the case that after the individual problem solving, the children form small groups and discuss their solutions and ideas. A variety of small groups are formed depending on the intention of the teacher. Sometimes four neighboring children come together to form a group and share different solutions proposed by the members of the group, choose the best solution and write it on the small whiteboard for the presentation. At other time, they rise from their desks and spontaneously form small groups, in which they teach each other how to solve the problem. To ensure that children engage in the activity of discussion, it is important to make clear both the purpose and object of discussion. It is also a challenge for the teacher to help every child to participate in the discussion.

Discussing solution methods. After the activity of solving the problem individually or in small groups, the teacher invites children to present their methods of solving the problem in front of the class. The teacher selects children in a particular order to encourage those children who found simpler methods. In some cases, even an incorrect method is presented. When the children present their solutions, the teacher prompts them to explain their thinking by using logical steps and mathematical terms, and by utilizing different tools and mathematical representations. The major purpose of presentation is to share children's ideas and solutions that will be compared and discussed next.

In the period of comparing and discussing different ideas and solutions, the teacher encourages children to become aware of the similarities and differences between their own ideas and those of their classmates'. Koto and a group of teachers proposed principles of organizing discussion (Koto & Niigata-ken-sansu-kyoiku-kenkyukai, 1992). They said that what should be done first is an examination of the validity of each solution. Then the relationships among different solutions should be discussed. How the multiple solutions are compared depends greatly on the nature of the problem and the objective of the lesson. Sometimes the class examines the relationship among solutions, such as how they are integrated or structured. At other times they order different solutions from the point of view of mathematics, such as, mathematical relevance, generality or utility. Lastly, they say that it is important that children look back and self-evaluate the solutions. The teacher supports and leads the discussion by asking children questions such as the following (see also Aguirre, 2012):

- Do you agree with Kaori's way of solving the problem?
- Do you understand Taro's explanation?
- Where is the same and what is different among these solutions?
- Which solution method is simpler?
- Which is the easiest to understand?
- Which solution method can be used for another problem?

It is especially important to ask children questions that attract their attentions to better ways of thinking mathematically.

Highlighting and summarizing the major points. In this stage the teacher leads children to reflect on what they have learned during the lesson and encourages them to write in their notebooks about their own ideas and their classmates' ideas that were discussed during the lesson. The teacher asks children, "What did you understand from the lesson today?" She/he tries to elicit children's thinking about their learning and use their phrases in writing conclusions on the blackboard. Shimizu (2006) says, "The solution methods of a problem or mathematical concepts and terms developed in the classroom are given recognition of validity and utility by the participants in the lesson, once they are summed up" (p. 142). The structured problem solving leads children to find alternative ways to solve a problem and so it is natural for the teacher to act in this way in the activity of summarising.

It should be noted that the roles of teacher described above are considered as the guidelines for the quality mathematics instruction via structured problem solving. We do not implement every lesson in this way. Nevertheless, especially when the objective of the lesson is to introduce a new mathematical concept or procedure to children, teachers make efforts to play these roles by trying to incorporate structured problem solving in their lessons.

3 Stimulating Reflective Thinking and Learning through Structured Problem Solving

3.1 *Perspectives on stimulating children's reflective thinking and learning*

It has been pointed out that Japanese lessons reflect certain values in the larger society and that one of such value is self-reflectivity (Lewis, 1996; Stigler & Hiebert, 1999; Shimizu, 2006, 2009a, 2009b). In this section, I first present some Japanese terms which are often discussed among teachers (see also Isoda, Stephens, Ohara, & Miyakawa, 2007). These terms provide perspectives on how we stimulate reflective thinking and learning through structured problem solving. This is followed by an examination of one textbook problem by using these perspectives.

"Kizuki" (noticing) and "Toi" (question). "Kizuki" or noticing is one of the keywords for stimulating children's reflection. A challenge of the teacher is how to motivate children to investigate and find the mathematical content which is the objective of the lesson. In planning and implementing a lesson, the teacher attempts to let children notice something new to them.

In this regard, it is considered important that children pose their own questions. The self-developed questions, which we call "Toi" in Japanese, are the driving force for children to think, discuss and notice something significant for them. Through exploring the questions, the children are also expected to engage in the activity of making connections. In order for children to notice important mathematics, teachers sometimes need to exercise leadership by picking out questions that are valuable to pursue and by suggesting the points of reflection. Eventually it is hoped that the children themselves ask good questions.

"Neriage" (kneading up). The second perspective is to knead up (or polish up) children's different solutions and ideas, which we call "Neriage" in Japanese. Reflection takes place naturally and substantially in the collaborative work with other people in the classroom. It is assumed that powerful individual learning becomes possible in the dynamics of learning as a classroom community. Lewis (1996) pointed out that Japanese teachers use groups to accomplish the dual goals of social and intellectual development. By working with their classmates in a whole-class activity, children are expected to learn better solutions and the mathematical concepts underlying these solutions (Sugiyama & Ito, 1990). To make it happen, teachers need to think about how to organize the order of presentations or the discussion after the presentations as described in the previous section.

"Yosa" (merit of mathematical ideas). The third perspective is the merit of mathematical ideas, which we call "Yosa" in Japanese. Long-standing research on mathematical thinking (e.g., Isoda & Katagiri, 2012) has offered an outlook on the values and advantages of mathematics. It is our premise that mathematics makes significant contributions to reflective thinking. Because of this, teachers plan to

teach mathematical concepts by providing the situation in which mathematical ideas behind the concepts become powerful means of reflection.

3.2 *An example: Compare the crowdedness of four rabbit cages*

In this sub-section, an example is given to show a way of incorporating the perspectives described above. By using a problem and descriptions in a textbook, in particular, I show how textbook writers also thoroughly examine the quality of the tasks used in the lessons.

Table 1
Area and number of rabbits

Cage	Area (m^2)	Number of rabbits
A	6	9
B	6	8
C	5	8
D	9	15

The tasks is, "There are four rabbit cages, A, B, C and D. Investigate how to order the four cages according to how crowded they are" (Fujii *et al.*, 2011). The purpose of this task for the fifth graders is to introduce per-unit quantity. To teach per-unit quantity, which is very important but difficult mathematical concept for children, we have been using the situation of comparing crowdedness among different locations. Points considered in design of the task include how many locations are compared and how many areas and the numbers of rabbits to be used in the task. Details such as the exact numbers to be used have been discussed since the numbers influence children's motivation to solve the task and, furthermore, reflect the generality of the task (Fujii, 2010). Table 1 shows the areas and the numbers of rabbits that were decided to be used to compare the four cages in the textbook. It might be a good activity for the readers to anticipate children's responses and to develop their plans of organizing different solutions.

Producing questions and opportunities for noticing. One natural way of comparing the four cages is "from easy ones to more difficult ones." A good way to begin, therefore, is by making comparisons between A and B, and between B and C. These comparisons are easier because in A and B the areas are the same, and in B and C the numbers of rabbits are the same. As a result, A is more crowded than B and C is more crowded than B. B is thus eliminated at this point. Children go to the next step of comparing A, C and D. However, this is more difficult and here is the opportunity for the children to develop their question, "In what way we can compare these cages in which neither areas nor numbers of rabbits are the same?" There is also the opportunity to connect what they have done and what they need to do and to notice the idea of changing different sizes to the same size.

Producing questions and opportunities for noticing by kneading up different solutions. In fact, it is possible to use different ideas to proceed to their activity of comparison. Since the children have compared two cages each time thus far, it is anticipated that many children try to compare two of the three cages in order, i.e., A and C, and C and D. In the textbook, three solutions are presented to compare A and C:

Solution 1

Make the area the same by using a common multiple of 5 and 6, or 30.

A ... $30 \div 6 = 5$, $9 \times 5 = 45$ (rabbits)

C ... $30 \div 5 = 6$, $8 \times 6 = 48$ (rabbits)

Therefore, C is more crowded than A.

Solution 2

Compare based on the numbers of rabbits in 1 m^2.

A ... $9 \div 6 = 1.5$ (rabbits)

C ... $8 \div 5 = 1.6$ (rabbits)

Therefore, C is more crowded than A.

Solution 3

Compare based on the amount of space for each rabbit.

A ... $6 \div 9 = 0.66\cdots$ (rabbits)

C ... $5 \div 8 = 0.625$ (rabbits)

Therefore, C is more crowded than A.

After the children share different ways of comparing the three cages in order, it is expected that questions emerge among children, for example, "Which idea is easier?" or "Which is better?" This is an opportunity to notice commonality and differences among the three solutions. The common part is to change different sizes to the same size to make the comparison possible. Different parts would include arithmetic operations used, numbers resulting from the calculations, or the levels of difficulty of interpreting the resulting numbers. The children can discuss their thinking on these matters. Associated with these questions, an important question that should be asked either by the children or by the teacher is, "Which idea is the easiest one to use when comparing the crowdedness of cages A, C and D altogether?" This question is worth examining because it will lead the children to the objective of the lesson.

Opportunities for noticing the merit of per-unit quantity. This question asks the children to compare different solutions from the viewpoint of generalizability. In other words, the question is, "Which solution or idea is most easily applied even when the number of cages increases?" Several anticipated responses are shown in the textbook:

Children1: Using a common multiple (Solution 1) becomes more difficult when you have to compare many cases.

Children2: If you make the area or the number of rabbits 1 (Solutions 2 and 3), even when there are many cases we can compare them easily.

These responses are directly related to the usefulness of finding the average number of rabbits in 1 m^2 or the average area for 1 rabbit. Therefore, this is an opportunity for the children to notice the merit of mathematical concept of per-unit quantity. Furthermore, the class can compare the easiness of Solutions 2 and 3. In Solution 2, the quantity becomes larger when it is more crowded. On the contrary, in Solution 3, the quantity becomes smaller when it is more crowded. On the basis of this difference, the children may agree that Solution 2 is easier to use for the purpose of judging the crowdedness from the per-unit quantity.

The above description indicates that questions are the driving force in the development of the lesson. Chains of questions repeatedly produce opportunities to gain insights and notice new mathematical ideas (Isoda *et al.*, 2007). The time for comparing and kneading up different ideas and solutions contributes greatly to the production of questions. Here again, it is worth mentioning that the numbers (Table 1) are chosen to direct the activity of comparison toward the moment at which the children notice the merit of mathematical concept of per-unit quantity. At the same time, four cages, not two or three cages, are decided on as the objects of comparison in this textbook. If there are just three cages A, B and C, children would compare A and C by using their own methods such as Solutions 1, 2 or 3. Opportunity to discuss generalizability might not occur unless the teacher is well aware of its significance. The presence of cage D is meaningful for the purpose of attracting children's attention to the generalizability and eventually, eliciting the power of per-unit quantity in the discussion.

4 Examples from Practice-Based Studies: Teachers' Trials in an In-Service Education Program

In this section, two practice-based studies are presented. Both studies were conducted by groups of teachers who participated in an in-service education program in Tochigi prefecture, Japan. In this program, the teachers were permitted to be away from their jobs continuously for the whole 6 month period and spend time for their professional development activities, in which they jointly discuss their tasks and conduct small-scale lesson study. Since the teachers come from different areas in the prefecture and different school levels (both primary and secondary schools), participation in this program offers them a unique opportunity to discuss and devise mathematics lessons collaboratively with members of a larger community.

4.1 *Providing a foothold*

In order to encourage children to develop their own questions, it is considered useful to provide a foothold when presenting today's problem to children. As described in the preceding section, children's self-developed questions offer important opportunities for them to reflect on their thinking and to connect it to the problem at hand. The objective of this lesson, which was for the fifth graders, was to think about the ways of finding an area of quadrilateral by decomposing or transforming the figure for which they had already leaned formulas (Kajino, Niimura, Kimura, Tanaka, & Kobayashi, 2009). In the lesson, the teacher let children find the area of a quadrilateral by applying the idea of finding the area of rhombus. Instead of giving the children one main problem, she prepared two problems before the main problem. Her intention was to encourage them to develop their interest as well as to reflect their thinking in the previous problem to develop a method for the next problem.

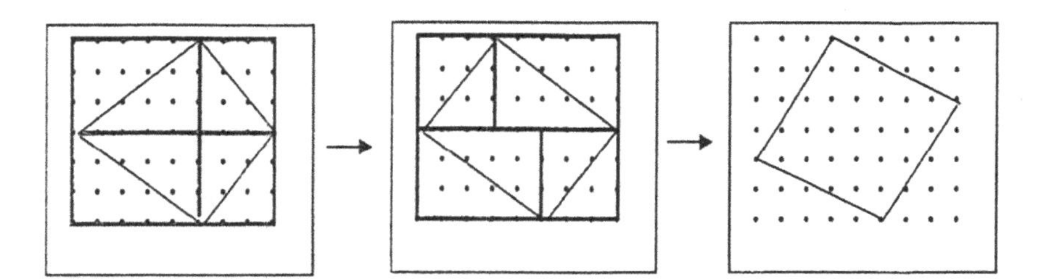

Figure 1. Two preceding tasks and today's task

The two figures and the main figure that were used in the classroom are shown in Figure 1. They were presented to children one by one. Each time the teacher asked the children, "Do you think the area of this figure is a half of the rectangle?" She framed the question in this way to motivate the children toward finding the area. For the first two figures, the children answered "Yes" eagerly. Several children made sure of their answers in front of the class. Lastly, the teacher presented the third figure, as the target of today's lesson, and asked the same question. At this time, the children were not sure whether it was a half or not. Their questions were triggered, which motivated them toward finding the area of this figure and checking whether it was a half of the area of the rectangle. Since the children had been thinking about the two previous quadrilaterals by applying the ways of finding the area of rhombus, e.g., decomposing the figure into different parts, or taking a half of the rectangle, they were motivated to apply these ideas when they developed ways of finding the area of the third figure.

After the lesson, the teacher surveyed the children about what they thought at different points of the lesson. One of the items was about this moment. The questionnaire item and the students' responses are given in Figure 2. Figure 2 shows that all of the children thought that they wanted to find the area of the third figure. Moreover, their writing shows that the children were motivated and also reflecting on their previous experience and learning.

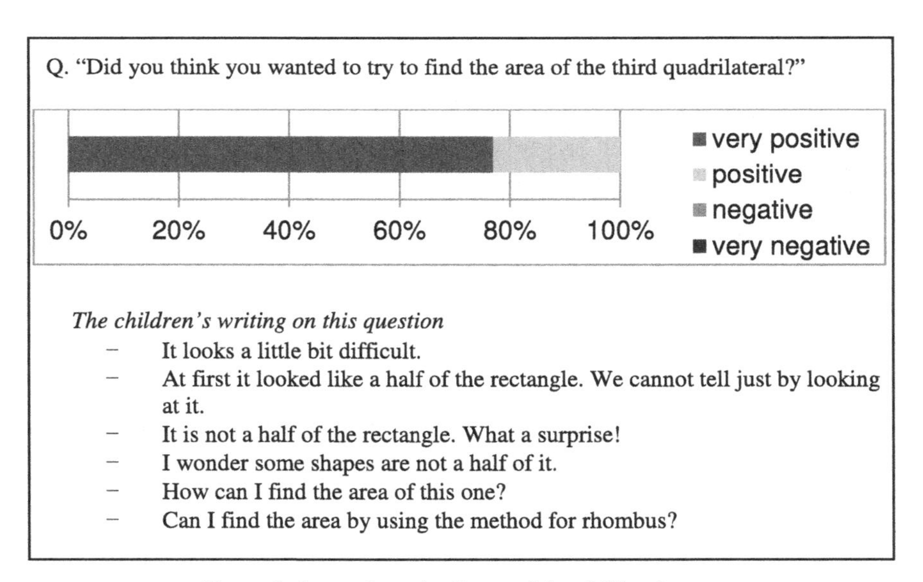

Figure 2. A questionnaire item and the children's responses

4.2 *Stimulating reflection through interaction*

In the latest version of CS, verbal activity during the lesson is stressed for the purpose of enhancing children's ability to think, make decisions, and to express their ideas to their classmates and the teacher (MEXT, 2008). Currently, we are accumulating teaching practices for verbal activities and we are becoming aware that good verbal activity includes rich opportunity for children to reflect on their thinking through the feedback from the listener and the speaker.

Designing and implementing the lesson. In this practice-based study, two teachers, one in primary school and the other in lower secondary school, developed lessons in which activities for learning by both generating and solving problems are incorporated (Sudo & Furuhashi, 2012). In the lessons in the fifth and the sixth grade classrooms, children were given the problem of finding the total number of marbles that were arranged in Figure 3. Before this problem, the teacher posed a simpler problem of marbles (marbles in a 2×6 array) and made clear what he intended them to do.

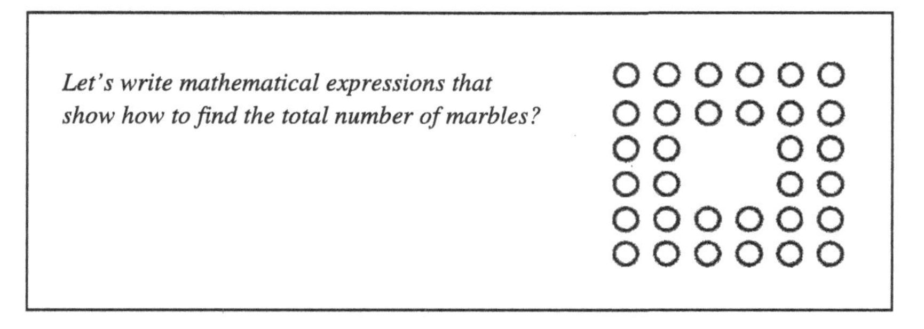

Figure 3. Task of counting marbles

Figure 4. Part of the worksheet for individual work

At first the children approached this task by using different mathematical expressions individually. They recorded their ways of counting marbles by drawing circles on the figures together with mathematical expressions on their worksheets (Figure 4). The mathematical expressions become the problems that the children posed to their partners in the next activity.

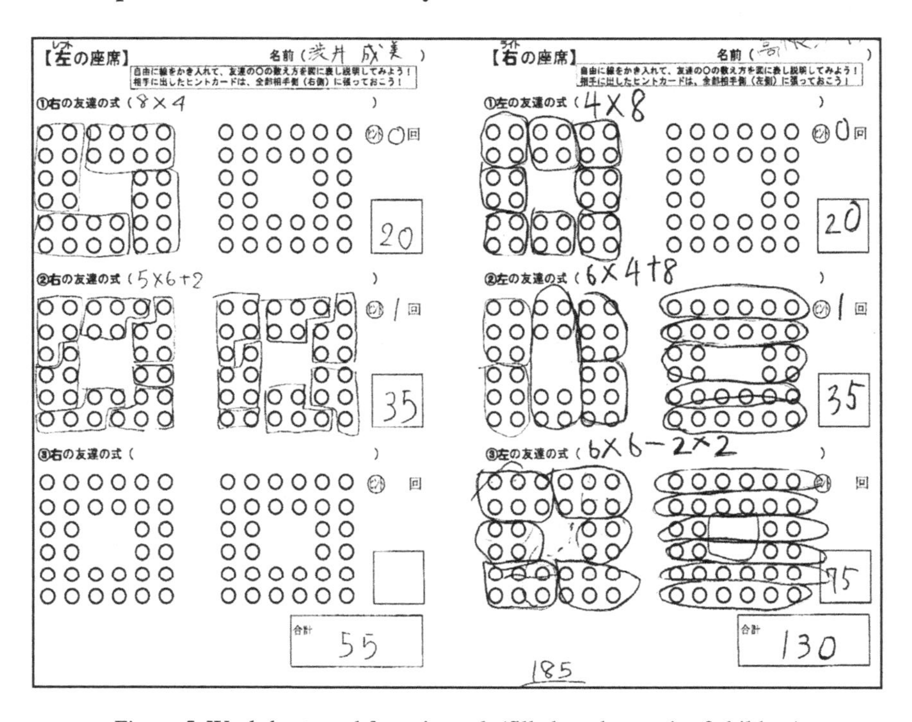

Figure 5. Worksheet used for pair work (filled out by a pair of children)

Next, in the pair work, the children who sat next to each other mutually presented the problems (mathematical expressions) to their partners and solved the problems (to find how the partners counted the marbles). In the worksheet that was distributed to each pair (Figure 5), one child (the examiner) presents one of her/his mathematical expressions to her/his partner (the respondent). The respondent reads the mathematical expression, conjecturing about the thinking behind the mathematical expression, and draws circles on the figure. If the figural representation does not match one of the examiner's, or if the respondent says, "I don't know," the examiner must give the respondent a clue. There are at most two chances to give the clue. After the pair work, the children did the same activity as a whole-class. Here, the teacher chose several mathematical expressions that were challenging for the children.

The worksheet used for the pair work was self-marked by each pair of children. The teacher distributed them a grading table (Figure 6). If the children solved more difficult problems, their scores were higher. If they solved the problem by asking for clues, they lost 5 points for each clue from the total score. Finally, each score was added as the total score for the pair of the children (see "185" in Figure 5) and total scores were presented later in class.

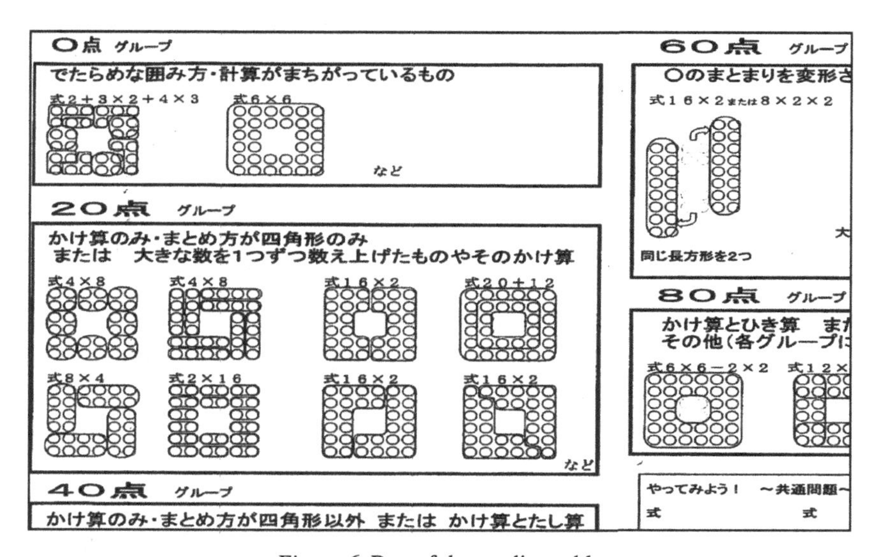

Figure 6. Part of the grading table

Evaluating and improving the lesson. After the lesson, the teacher summarized information obtained from observations, field notes, worksheets and a video camera together with comments made by the observers. He reported his findings about the children's behaviors as well as about the points for improving his teaching (Sudo & Furuhashi, 2012), which are described below.

When the children were asked to solve the simpler problem, the teacher let them make conjectures about how to count the marbles. They worked in pairs for 1 or 2 minutes. Although the time for interaction was short, they compared their thinking with their friends and began to notice that there were a variety of ways of counting the marbles. The teacher found this short exchange was important for the children to engage in the task productively.

When the children were working individually, the teacher walked around their desks and checked their progress. He recognized that the children tended to make difficult problems or many problems from the beginning. As a result, he reported that it is important to say, "Let's make problems from easier ones" or "Let's make one problem first," so that they can make progress at their own pace. He also reported that it is important to verbalize the ideas that he found among the children, e.g., "Oh, I knew that we can use subtraction, too," so that they can extend their ideas.

One core part of this lesson is the pair work. Here, the children need to think, reflect and decide their behaviors in response to the feedback from their partners. In particular, the child needs to give clues to her/his partner. For the purpose of attracting the children's attention to this important action, the teacher asked them not only to speak but also to write the clues on the post-it cards. Using the post-it cards for this purpose was proposed from the repeated discussion among the teachers. The clues collected with the worksheets show that the children anticipated the thinking of their partners, recognized the discrepancies between their own and their partner's solutions and modified their clues to more precise ones. For example, one child first wrote "Divide the total in the group of 5 marbles." However, this clue did not offer useful information because the mathematical expression already included "5×6." Later, the other child wrote, "Long and skinny" and "Find the total area." This pair of children was attending to the shapes and methods, or wider perspectives, in reading mathematical expressions. The teacher reported that generating clues enhance the children's understanding of ideas behind solutions.

After the pair work, the children read several mathematical expressions as a whole-class activity. The teacher reported that in order to introduce mathematical expressions that are valuable to the children mathematically, he needed to examine anticipated mathematical expressions from the point of view of connectivity and integrity.

The grading table was also developed after the repeated discussion among the teachers. The intent was that the children could attend to different ideas more explicitly by grading their solutions. At the same time, it was considered important to calculate the total score as the score of *the pair*. This requirement motivated the children not to compete against each other but to work collaboratively as a pair. As a result, the children took care in the thinking of their partners when they posed problems and gave them clues.

5 Conclusions

In this chapter, the Japanese problem solving approach and several perspectives for enhancing reflective thinking and learning during the classroom teaching have been described. The problem solving approach includes a powerful methodology for enhancing reflection by providing children with opportunities to develop their own questions and to investigate, discuss and notice significant mathematics for them. However, to realize these opportunities, the teacher plays an essential role. The teacher needs to support children to propose the points and methods of reflection. It is also important that the teacher supports children to connect the results of their reflections to new problems. Children need to experience both the process and the advantage of reflection in their learning of mathematics.

In the latter part of the chapter, some of teachers' trials for encouraging reflective thinking in the lesson were illustrated. These illustrations suggest that there are a variety of ways to approach our aim of encouraging children's reflective thinking and learning. This is in line with the challenge of crafting a variety of high-level cognitive demand tasks to advance reasoning and communication from closed textbook tasks (e.g., Kaur & Yeap, 2009a, 2009b; Thompson, 2012). The style of structured problem solving *per se* is not the solution. Rather, designing, implementing, evaluating and improving such approaches in the lessons contribute to the professional development of teachers. This is because it enhances teachers' awareness that reflective thinking is so fundamental in children's learning of mathematics and that developing such attitudes among children empowers them in their future lives.

References

Aguirre, K. (2012). Comparison of mathematics teaching methodology in first and second grade of elementary education between Japan and Peru. *Utsunomiya University International In-Service Teacher Training Program Report* (pp. 55-74). Graduate School of Education, Utsunomiya University.

Fujii, T. (2010). Designing tasks in the Japanese lesson study: Focusing on the role of the quasi-variable. In Y. Shimizu, Y. Sekiguchi, & K. Hino (Eds.), *Proceedings of the 5th East Asia Regional Conference on Mathematics Education* (Vol. 1, pp. 86-93). Tokyo, Japan: EARCOME.

Fujii, T. *et al.* (2011). *New mathematics grade 5.* (Mathematics 1-6 set, DVD-ROM. English translation of elementary school textbook) Tokyo: Tokyo Shoseki.

Hiebert, J., Gallimore, R., Garnier, H., Givvin, K.B., Hollingsworth, H., Jacobs, J., Chiu, A.M.-Y., Wearne, D., Smith, M., Kersting, N., Manaster, A., Tseng, E., Etterbeek, W., Manaster, C., Gonzales, P., & Stigler, J. (2003). *Teaching mathematics in seven countries: Results from the TIMSS 1999 Video Study.* U.S. Department of Education. Washington, DC: National Center for Education Statistics.

Hino, K. (2007). Toward the problem-centered classroom: Trends in mathematical problem solving in Japan. *ZDM – International Journal on Mathematics Education, 39*, 503-514.

Hino, K. (2009). *Coherence in student's construction of mathematical meanings: Glimpses from three Japanese classrooms.* Paper presented at the 3rd Redesigning Pedagogy International Conference, National Institute of Education, Singapore, June 1-3.

Isoda, M., & Katagiri, S. (2012). *Mathematical thinking: How to develop it in the classroom.* Singapore: World Scientific.

Isoda, M., Stephens, M., Ohara, Y., & Miyakawa, T. (Eds.). (2007). *Japanese lesson study in mathematics: Its impact, diversity and potential for educational improvement*. London: World Scientific.

Kajino, Y., Niimura, K., Kimura, S., Tanaka, Y., & Kobayashi, T. (2009). Sansu/sugakuteki katsudo wo toriireta jyugyo no kufu [Incorporating mathematical activity into the lesson]. *Report of the in-service teacher training, Tochigi Prefecture.*

Kaur, B., & Yeap, B.H. (2009a). *Pathways to reasoning and communication in the primary school mathematics classroom*. Singapore: National Institute of Education.

Kaur, B., & Yeap, B.H. (2009b). *Pathways to reasoning and communication in the secondary school mathematics classroom*. Singapore: National Institute of Education.

Koto, S., & Niigata-ken-sansu-kyoiku-kenkyukai (Eds.). (1992). *Sansuka tayo na kangae no ikashikata matomekata [Ways of utilizing and summarizing various ways of thinking in elementary mathematics class]*. Tokyo: Toyokan.

Lewis, C. (1996). Fostering social and intellectual development: The roots of Japanese educational success. In T. Rohlen & G. LeTendre (Eds.), *Teaching and learning in Japan* (pp. 79-97). New York: Cambridge University Press.

MEXT. (2008). *Shogakko gakushu sido yoryo kaisetsu: Sansu hen [Elementary School Teaching Guide for the Japanese Course of Study: Mathematics]* (English translation was carried out by the Asia-Pacific Mathematics and Science Education Collaborative at DePaul University in Chicago, Illinois, U.S.A., under contract from the U.S. Department of Education)

Sekiguchi, Y. (2012). An analysis of coherence and variation in Japanese mathematics classrooms. *Pre-proceedings of the 12th International Congress on Mathematical Education* (pp. 4332-4341). Seoul, Korea: ICME.

Shimizu, Y. (1999). Aspects of mathematics teacher education in Japan: Focusing on teachers' roles. *Journal of Mathematics Teacher Education 2*, 107-116.

Shimizu, Y. (2006). How do you conclude today's lesson?: The form and functions of 'Matome' in mathematics lessons. In D. Clarke, E. Jonas, E. Jablonka, & I.A.C. Mok (Eds.), *Making connections: Comparing mathematics classrooms around the world* (pp. 127-145). Rotterdam: Sense Publishers.

Shimizu, Y. (2009a). Characterizing exemplary mathematics instruction in Japanese classrooms from the learner's perspective. *ZDM – International Journal on Mathematics Education, 41*, 311-318.

Shimizu, Y. (2009b). Japanese approach to teaching mathematics via problem solving. In B. Kaur, B.H. Yeap, & M. Kapur (Eds.), *Mathematical problem solving: Yearbook 2009* (pp. 89-101). Singapore: World Scientific.

Stigler, J.W., & Hiebert, J. (1999). *The teaching gap*. New York: Free Press.

Sudo, H., & Furuhashi, N. (2012). Jibun no kangae wo setsumei shi tsutaeau katsudo wo jyushi shita jyugyo no kufu [Mathematics lesson for the activity of explaining and communicating one's thinking with other people]. Report of the in-service teacher training, Tochigi Prefecture.

Sugiyama, Y., & Ito, S. (Eds.). (1990). *Sansuka jyugyo kenkyu [Lesson study in elementary mathematics]*. Tokyo: Kyoiku Shuppan.

Thompson, D.R. (2012). Modifying textbook exercises to incorporate reasoning and communication into the primary mathematics classroom. In B. Kaur & T.L. Toh (Eds.), *Reasoning, communication and connections in mathematics: Yearbook 2012* (pp. 57-74). Singapore: World Scientific.

Chapter 14

Insights from Students' Process of Understanding Mathematics for Nurturing Reflective Learners

Masataka KOYAMA

It is well recognized that reflective thinking is key for students to learn mathematics and deepen their mathematical understanding. Thus, we need to look into students' process of understanding mathematics in a classroom for insights that may help us nurture reflective learners. In this chapter we focus on a 8th grade lesson on "regular stellar polygon" in a Japanese lower secondary school classroom, and analyze the students' process of understanding using the so-called "two-axis process model" (Koyama, 1993, 1997a, 1997b). The outcome is four suggestions for nurturing reflective learners in lower secondary school mathematics. Firstly, teachers should pay attention not only to the correctness of student answers to mathematical tasks but also to the processes that lead to the answers. Secondly, it is important for teachers to make a decision on the kinds of learning situation that need to be set up so as to help students improve their mathematical understanding. Thirdly, the social interaction between teacher and students is important to ensure a high level of mathematical understanding. Lastly, students should be encouraged to reflect on what they have done and also do adequate activities to integrate their knowledge.

1 Introduction

It is well recognized that reflective thinking is key for students to learn mathematics and deepen their mathematical understanding (van Hiele & van Hiele-Geldof, 1958; Pirie & Kieren, 1989; Nakahara, 1995; Koyama, 1995a, 1995b). For nurturing reflective learners in mathematics, we need to look into students' process of understanding mathematics in a classroom.

As a descriptive and prescriptive model, Koyama (1993, 1997a, 1997b) made the so-called "two-axis process model" of mathematical understanding. The model consists of the vertical axis comprising three levels of understanding such as mathematical entities, relation of the entities and general relation, and the horizontal axis comprising three learning stages of intuitive, reflective, and analytic at each level. As a result of theoretical and practical studies, Koyama (2005) identified the principles and methods for designing mathematics lessons based on the "two-axis process model" of mathematical understanding. Through a series of case studies in primary school mathematics, it has been demonstrated that the model can be used by teachers as an effective framework for designing mathematics lessons to improve their students' mathematical understanding in a classroom (Koyama, 2007, 2010).

However, we need more case studies on the students' process of understanding in secondary school mathematics. In this chapter we focus on a 8th grade lesson on "regular stellar polygon" in a Japanese lower secondary school classroom, and analyze the students' process of understanding mathematics using the "two-axis process model". The purpose of this qualitative analysis is to get some suggestions for nurturing reflective learners in lower secondary school mathematics.

2 The "Two-Axis Process Model" of Mathematical Understanding

In previous studies on mathematical understanding, two types of model are proposed. One is the *aspect model* found in Skemp (1982) and Byers and Herscovics (1977), and the other is the *process model* found in Herscovics and Bergeron (1988), and Pirie and Kieren (1989, 1994). These models are distinct. The former is used to identify different kinds of mathematical understanding at a particular moment, and the latter describes a process of mathematical understanding.

Both these models have a *descriptive characteristic* in that they can describe the kinds or the process of mathematical understanding in learning mathematics. However, the descriptive aspect alone is not enough for us to design a mathematics lesson because mathematics lessons must be organized as integrated activities of teaching and learning. Therefore we need a model with not only a descriptive characteristic but also a *prescriptive characteristic* that can suggest some didactical principles such as what kind of problematic situations should be set up and which direction should be aimed at for improving students' mathematical understanding in the process of teaching and learning activities in a classroom (Koyama, 1992).

Based on theoretical studies to identify basic components for a descriptive and prescriptive model of mathematical understanding (van Hiele & van Hiele-Geldof, 1958; Wittmann, 1981; Pirie & Kieren, 1989), Koyama (1993, 1997a, 1997b) made the so-called "two-axis process model" of mathematical understanding. The model consists of two axes, the vertical and the horizontal. The vertical axis comprises three levels of understanding mathematics and the horizontal axis comprises three stages of learning at each level. The levels of understanding mathematics and stages of learning are as follows:

Vertical axis

V1: mathematical entities

V2: relation of the entities

V3: general relation

Horizontal axis

H1: Intuitive stage

Students are provided opportunities for manipulating concrete objects, or operating on mathematical concepts and relations acquired in a previous level. At this stage, they do *intuitive thinking*.

H2: Reflective stage

Students are stimulated and encouraged to pay attention to their own manipulating or operating activities, to be aware of them and their consequences, and to represent them in terms of diagrams, figures or language. At this stage, they do *reflective thinking*.

H3: Analytic stage

Students elaborate their representations to be mathematical ones using mathematical terms, verify the consequences by means of other examples or cases, or analyze the relations among consequences in order to integrate them as a whole. At this stage, they do *analytic thinking*.

Based on theoretical and practical studies, Koyama (2005) identified three principles and three methods for designing mathematics lessons based on the "two-axis process model" to facilitate and deepen students' mathematical understanding as follows:

Principles for designing mathematics lessons

P1: recognizing mathematical understanding as a dynamic process

P2: setting up levels of understanding and learning stages on a level

P3: incorporating students' individual constructions and social constructions

Methods for designing mathematics lessons

M1: making clear levels of understanding related to a certain mathematical topic

M2: assessing and evaluating students' understanding as a readiness

M3: planning in detail three learning stages as a dialectic process of individual and social constructions in a lesson

Through a series of case studies in primary school mathematics, it has been demonstrated that the "two-axis process model" and the principles and methods can be used by teachers as an effective general framework for designing mathematics lessons to improve their students' mathematical understanding in a classroom (Koyama, 2007, 2010).

However, we need more case studies on the students' process of understanding in secondary school mathematics. In the following sections, first we focus on a 8th grade lesson on "regular stellar polygon" in a Japanese lower secondary school classroom. Then we analyze the students' process of understanding mathematics in the lesson using the "two-axis process model". Finally we summarize some suggestions for nurturing reflective learners in lower secondary school mathematics.

3 Development of a 8th Grade Lesson on "Regular Stellar Polygon"

3.1 *Objectives of the lesson*

The lesson to be analyzed is a fifty-minute lesson on "regular stellar polygon" with 39 Grade 8 students in a lower secondary school attached to Hiroshima University. This lesson was designed by a mathematics teacher, Mr. Tominaga, as a task-based learning lesson as part of the teaching unit of "triangles and quadrangles" for 8th graders on 14th November 2003. Prior knowledge of students comprised properties of parallel lines, sums of interior angles and exterior angles of a polygon, properties of an isosceles triangle, the theorem of angles at the circumference, and so on as prescribed in the Japanese Course of Study for lower secondary school mathematics (Ministry of Education, Science, Sports, Culture and Technology, 1999). The objectives of the lesson on "regular stellar polygon" were to improve students' understanding of how to find out the measure of an angle in a regular stellar polygon (see Figure 1) and to promote their mathematical thinking and attitude towards mathematics through generalizing the mathematical tasks on regular polygons to regular stellar polygons (Tominaga, 2003).

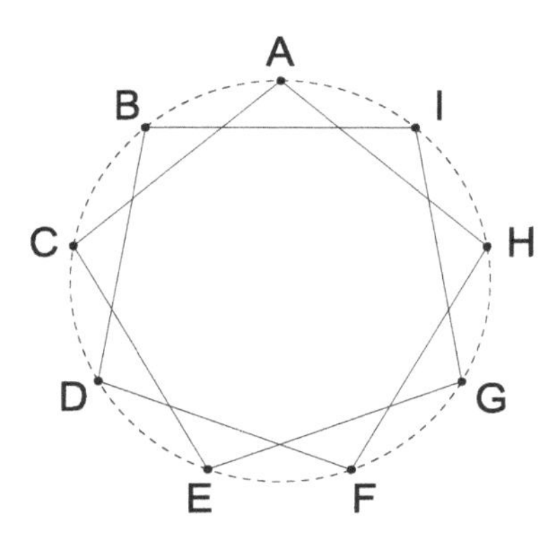

Figure 1. A regular nine-pointed stellar polygon for $k = 2$

3.2 *A brief of the lesson*

The lesson was conducted as a part of open lesson study. The author collected the data of the lesson by observation, note taking and video recording. The lesson developed as follows.

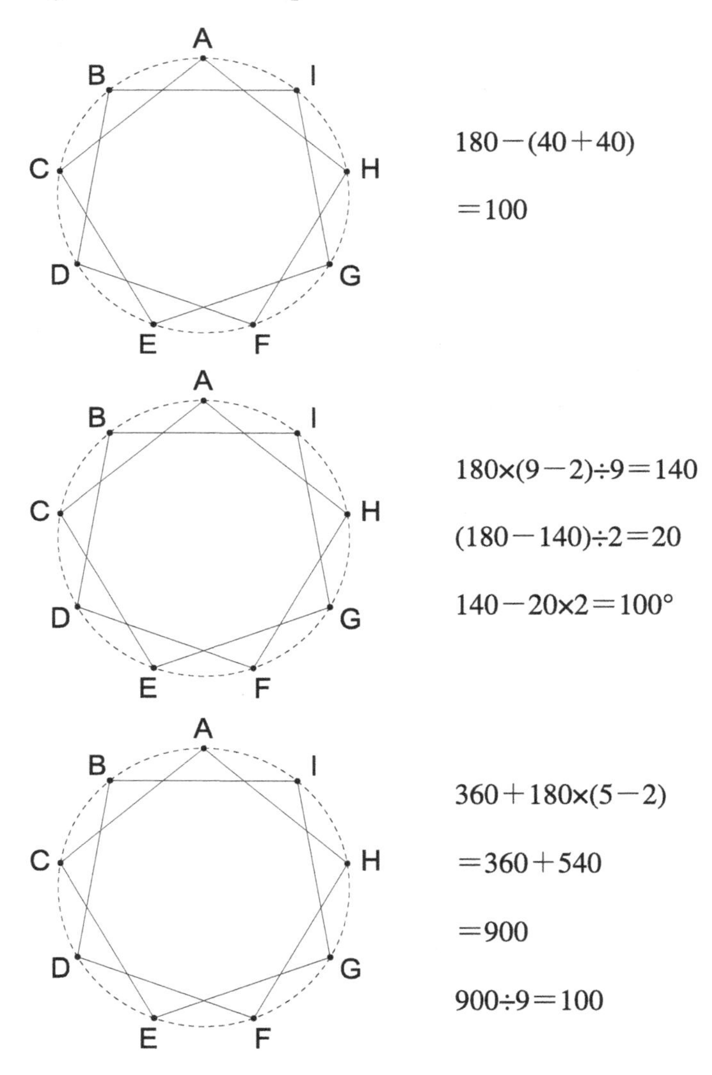

Figure 2. Solutions of three students called on by teacher

In the introduction part of the lesson, after reviewing how to draw a regular stellar polygon, the teacher posed his students the first task, i.e. task 1 "Find out the measure of a tip angle in the regular nine-pointed stellar polygon made by connecting next two points as shown in Figure 1". During students' individual work on the task, the teacher did a round of checking (*kikan-shido*). He noticed three different types of students' solution, and he called on three students to write their own solutions on a blackboard (see Figure 2). Then the teacher asked each student to explain his/her solution as follows.

S1: (looking at the teacher in a low voice) I do not need to say that this (△AJK) is an isosceles triangle to my classmates (see Figure 3)?

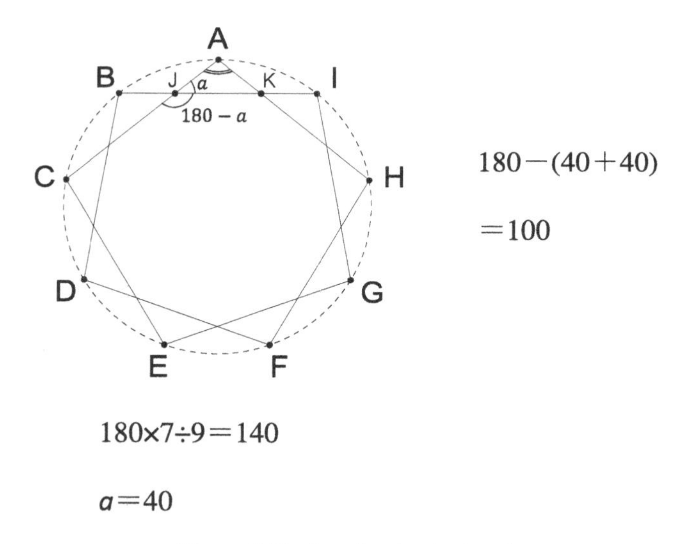

Figure 3. Explanation by Student 1

T: What? This is an isosceles triangle? You need to say clearly it to your classmates. I must move to your classmates.

S1: Yes. This triangle (△AJK) is an isosceles triangle...
So let this (∠AJK) is $\angle a$, then here (∠IJC) is $180 - a$. The measure of one angle of the nine-cornered polygon (nonagon) inside a circle is $180 \times 7 \div 9$, $140°$. So $180 - 140$, $\angle a = 40°$. As I said, because this triangle (△AJK) is an isosceles triangle,

here (∠AKJ) is also $\angle a$, $40°$. Because the sum of interior angles of a triangle is $180°$, 180 minus 40 and 40, the answer is $100°$.

S2: The first expression means that the measure of one angle of the nine-cornered polygon (nonagon) inscribed in a circle, looking at this triangle (△AHI), here (∠AIH) is 140 (see Figure 4).

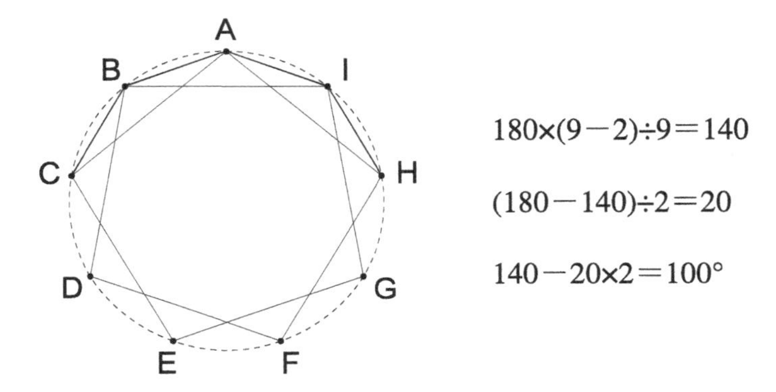

Figure 4. Explanation by Student 2

T: It is the same explanation as S1's. The nine-cornered polygon...

S2: Then, because the orange colored triangle (△AHI) is an isosceles triangle, here (∠IAH) and here (∠IHA) are $20°$. This triangle (△ABI) is same as that triangle (△AHI). So here (∠AIB) and here (∠ABI) are $20°$. Then $140°$ minus $2\times 20°$ is $100°$.

S3: (After drawing parallel lines AB and CI as shown in Figure 5) Looking at this small triangle (△ABJ) and this triangle (△CIJ), because these are vertically opposite angles here (∠AJB and ∠CJI) is same. Um... When this side (∠JIC and ∠JCI) is moved to that side (∠JAB and ∠JBA), we have a pentagon ABDFH. Then we have a quadrangle CEGI. So the sum of all interior angles in a pentagon ABDFH and a quadrangle CEGI is $540°$ plus $360°$. It is $900°$. Because all tip angles are equal, $900°$ divided by 9 is $100°$.

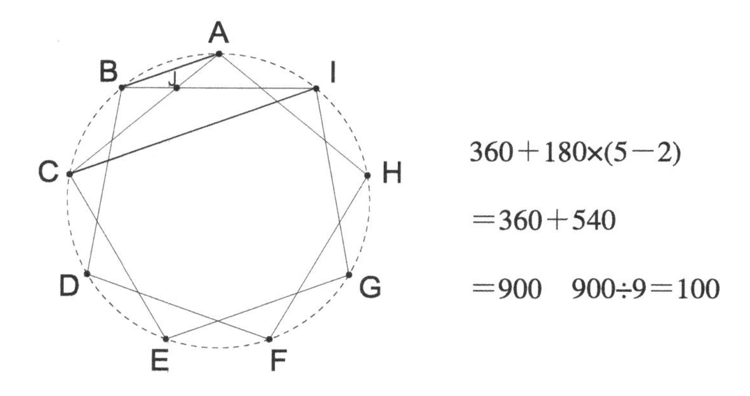

Figure 5. Explanation by Student 3

Following the three students' explanations, the teacher said "There are different solutions leading to the same answer 100°. Do you agree?" There was no voice from the students. Then the teacher posed his students the second task, i.e. task 2 "Investigate the measure of a tip angle in the regular nine-pointed stellar polygon made by connecting next k points such that $k = 1,2,3,4$, and state your findings".

He allowed his students to work together on the task if they needed. During students' work on the task, the teacher did a round of checking (*kikan-shido*), and confirmed that in the case of $k = 1$ the regular nine-pointed stellar polygon is a regular nonagon and the measure of a tip angle is $140°$, and that when $k = 3$ the regular nine-pointed stellar polygon is an equilateral triangle and the measure of a tip angle is $60°$. Then he said "We have a problem when $k = 4$. What a complex figure this is! Answer the measure of a tip angle in this case" (see Figure 6).

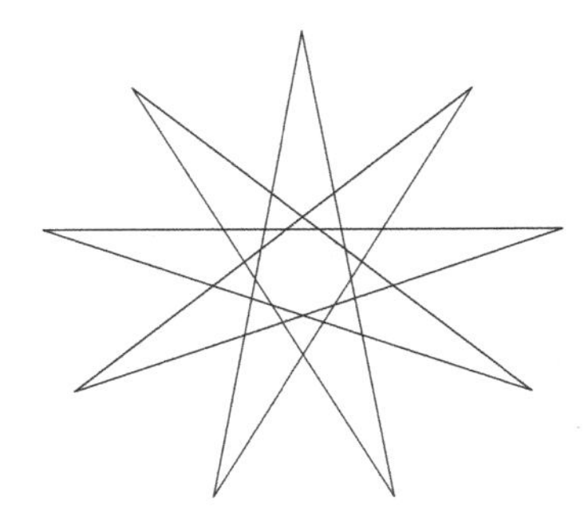

Figure 6. The regular nine-pointed stellar polygon for $k = 4$

He called on a student (S4) to explain her solution. The student S4 tried to explain the measure of a tip angle is $20°$ by using the theorem of angles at the circumference as follows (see Figure 7).

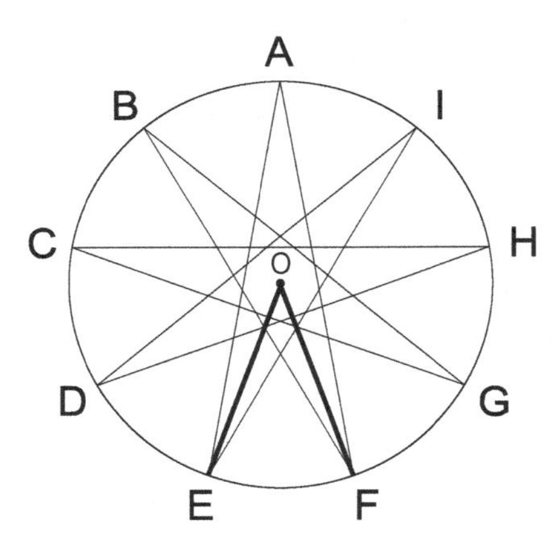

Figure 7. Using the theorem of angles at the circumference for $k = 4$

S4: If we draw a circumscribed circle in the Figure 6, a central angle of this is $360°$ divided by 9, $40°$.

T: I see. The central angle is $40°$. So?

S4: Um... The angle at the circumference is... (Her voice was not heard.)

T: Yes, you are right. Because the angle at the circumference is half of a central angle, the angle at the circumference in this case is $40°$ divided by 2, $20°$. Ok, you are right.

The teacher summarized the answers for $k = 1,2,3,4$ into a table (see Figure 8) and asked his students a question "Look at this table. What can you find out?"

k	1	2	3	4
the measure of a tip angle	140	100	60	20

Figure 8. The table made by the teacher

A student (S5) answered that the measure of a tip angle decreases at the rate of $40°$ as the value of k increases one by one. The teacher approved the finding of student S5 and prompted others.

S6: Well, $20°$ is a unit, $60°$ is three times, $100°$ is five times, $140°$ is seven times...

T: I see. You say that $140°$ is seven times of $20°$, $100°$ is five times of $20°$, $60°$ is three times of $20°$, and $20°$ is one times of $20°$. So one, three, five, and seven are odd numbers. Ok, this is an excellent property. Are there any other findings?

SS: (There was no reaction from the students.)

T: Well, you said that the measure of a tip angle decreases at the rate of $40°$. Why $40°$? The fact that the measure of a tip angle decreases at the rate of $40°$ is valid only for a regular nine-polygon? How about for a regular eight-polygon, ten-polygon, six-polygon, 12-polygon, 24-polygon, and so on? Anyway, why $40°$ for a regular nine-polygon?

S7: The central angle?

T: The central angle! The central angle is $40°$. It seems that there is a relation between the tip angle and the central angle. We have no time to explore the relation now. Let's think about the task 3 "Represent the measure of a tip angle in the regular

n-pointed stellar polygon made by connecting next k points in an algebraic expression with letters n and k".

At this point the lesson period was over. There was insufficient time for the students to attempt task 3 and therefore the task was left for the next mathematics lesson.

4 Qualitative Analysis of Students' Process of Mathematical Understanding in the Lesson

4.1 *Three tasks posed by the teacher*

The objectives of the lesson were to improve students' understanding of how to find out the measure of an angle in a regular stellar polygon (see Figure 1) and to promote their mathematical thinking and attitudes towards mathematics through generalizing the mathematical tasks on regular polygons to regular stellar polygons. In this lesson the teacher posed three tasks which were as follows:

Task 1: "Find out the measure of a tip angle in the regular nine-pointed stellar polygon made by connecting next two points as shown in Figure 1".

Task 2: "Investigate the measure of a tip angle in the regular nine-pointed stellar polygon made by connecting next k points for $k = 1,2,3,4$, and state your findings".

Task 3: "Represent the measure of a tip angle in the regular n-pointed stellar polygon made by connecting next k points in an algebraic expression with letters n and k".

In the lesson the students worked only on tasks 1 and 2. Task 3 was left for the next lesson. Therefore, we analyze students' process of mathematical understanding by studying how they attempted tasks 1 and 2 using the "two-axis process model".

4.2 *Solution process of task 1*

The first task, i.e. task 1 is at the level V1 (mathematical entities) in the vertical axis of the "two-axis process model" because the measure of a tip angle in the regular nine-pointed stellar polygon made by connecting next two points is the object in this task. At least three students S1, S2, and S3 could reason logically and explain that the measure of a tip angle is $100°$ by using their understandings already learned in previous mathematics lessons without any teacher's explicit setting of the three learning stages H1, H2, and H3 in the horizontal axis. However, the points aimed at by these students were different as follows:

S1: The triangle AJK is an isosceles triangle (see Figure 3).

S2: The triangle AHI is an isosceles triangle (see Figure 4).

S3: The regular nine-pointed stellar polygon made by connecting next two points is transformed into a quadrangle CEGI and a pentagon ABDFH without changing the sum of nine tip angles (see Figure 5).

Although the two students S1 and S2 focused on different isosceles triangles, they found out that the measure of a tip angle is $100°$ by their logical reasoning on the basis of the facts that two base angles of an isosceles triangle are equal and that the measure of an interior angle of a regular nine-polygon is $140°$. In contrast student S3 transformed logically the regular nine-pointed stellar polygon into a quadrangle and a pentagon without changing the sum of nine tip angles, and found that the measure of a tip angle is $100°$ by the calculation of $(360° + 540°) \div 9$. From this observation, we see that these three students attained the level V1 of understanding mathematical entities in task 1 and that their solution methods were different. This suggests that in a mathematics lesson when a teacher evaluates his/her student's mathematical understanding he/she should pay attention not only to the correctness of student's answer to a task but also to the processes that lead to the answer.

4.3 *Solution process of task 2*

The second task, i.e. task 2 is at level V2 (relation of mathematical entities) in the vertical axis of the "two-axis process model" because the relation among tip angles in the regular nine-pointed stellar polygon made by connecting next k points ($k = 1,2,3,4$) is the object in this task.

It means that task 2 intends to improve the students' mathematical understanding from level V1 to level V2. When the students worked on this more complicated task 2 in a regular nine-pointed stellar polygon, first they tried to draw the regular nine-pointed stellar polygons in each case of $k = 1,2,3,4$. At this point the object of the students' understanding was temporary shifted from investigating a relation among the measures of tip angles to having the images of regular nine-pointed stellar polygons by connecting the next k points ($k = 1,2,3,4$). This phenomenon is known as the folding back from the property-noticing level to the image-making level in the transcendent recursive model of mathematical understanding (Pirie & Kieren, 1989).

When working on task 2, the phenomenon of folding back allowed students to go back to the images of regular nine-pointed stellar polygons ($k = 1,2,3,4$) and formalize the relation among the measures of tip angles of regular nine-pointed stellar polygons ($k = 1,2,3,4$).This shows the importance of teacher's decision making related to the kinds of learning situation that can be set up so that students improve their mathematics understanding in a lesson. If the teacher intends to improve the students' mathematical understanding on the level V2 (relation of the entities) in the "two-axis process model", then he should set up learning situations by giving careful consideration to three stages of H1, H2, and H3. For example, if the following alternative tasks were posed by the teacher, the students could be stimulated and encouraged to pay attention to their own manipulating activities, to be aware of the consequences, and to represent them in terms of geometrical figures and mathematical language and expressions. The task 2-1 is a task for the intuitive stage and the task 2-2 is for the reflective and analytic stages on the revel of the relation of mathematical entities.

Task 2-1: "Draw the regular nine-pointed stellar polygons made by connecting next k points for $k = 1,2,3,4$".

Task 2-2: "Find out and make a table of the measures of tip angle in the regular nine-pointed stellar polygon made by connecting next k points for $k = 1,2,3,4$".

It is important for a teacher to set learning objectives for his/her students in a mathematics lesson. Also for the same learning objectives varied learning situations may be set up in a lesson. Therefore the teacher must investigate the actual situation of students' mathematical understanding, and make a decision on the kinds of learning situation to be set up for helping students improve their mathematics understanding. For this purpose, the "two-axis process model" with its principles and methods can be used by the teacher as an effective general framework for designing a mathematics lesson to improve the students' mathematical understanding in a classroom.

4.4 *Transition from task 2 to task 3*

In the lesson, the teacher summarized the answers for $k = 1,2,3,4$ into a table (see Figure 8) and asked his students a question "Look at this table. What can you find out?" A student (S5) answered that the measure of a tip angle decreases at the rate of $40°$ as the value of k increases by one. Another student (S6) explained that $20°$ is a unit for the measures of a tip angle. At this moment, the students did not connect these findings with the fact that the original figure was a regular nine-polygon. Therefore the teacher asked "Why $40°$? The fact that the measure of a tip angle decreases at the rate of $40°$ is valid only for a regular nine-polygon?" to all students. Here we can see an intention of the teacher in his questioning to deepen and direct the students' mathematical understanding of a relation among the measures of a tip angle in the regular nine-pointed stellar polygon made by connecting next k points for $k = 1,2,3,4$". However, we could not hear any clear voices from the students. Only one student (S7) answered in a mutter "The central angle?" to the question. Although his answer is insufficient, we guess that he could notice the central angle because he carefully heard when the student (S4) tried to explain the measure of a tip angle is $20°$ (in case of $k = 4$) by using the theorem of angles at the circumference (see Figure 7). This shows that the social interaction between teacher and students in a mathematics classroom is important in order to develop the lesson to a higher level of mathematical understanding.

In this lesson, the teacher hurried to the next task, i.e. task 3 "Represent the measure of a tip angle in the regular n-pointed stellar polygon made by connecting next k points in an algebraic expression with letters n and k". It was a key point for the teacher's decision making in the lesson. At this point it would have been better for the teacher to encourage the students to make a connection amongst the measures of a tip angle in a regular nine-pointed stellar polygon made by connecting next k points for $k = 1,2,3,4$. Task 3 was at level V3 (general relation) in the vertical axis of the "two-axis process model". Therefore, students should be provided more time and opportunities to reflect on what they have done in task 2 (reflective stage) and also do more activities for integrating them (analytic stage) according to the learning stages in the "two-axis process model" of mathematical understanding before proceeding to task 3.

5 Conclusion

It is well recognized that reflective thinking is key for students to learn mathematics and deepen their mathematical understanding. For nurturing reflective learners, we need to look into students' process of understanding mathematics in a classroom. In this chapter, we focused on a 8th grade lesson on "regular stellar polygon" in a Japanese lower secondary school classroom, and analyzed the students' process of understanding using the so-called "two-axis process model".

The qualitative analysis of the lesson resulted in four suggestions for nurturing reflective learners in lower secondary school mathematics. First, there were three different approaches to task 1 of a regular nine-pointed stellar polygon. This shows that the teacher should pay attention not only to the correctness of students' answer to a task but also to the process lead to the answer. Second, when the students worked on the more complicated task 2 of a regular nine-pointed stellar polygon they folded back to the image-making level. This demonstrates the importance of teacher's decision making on the kinds of learning situation to be set up for helping students improve their mathematics understanding. Third, the teacher's questioning about the reason and respective student's response changed the development of the lesson to a higher level of understanding. This shows that the social interaction between teacher and students is important in order to develop a lesson to a higher level of mathematical understanding. Finally, the students had difficulties when they worked on the more generalized task, i.e. task 3, for a regular n-pointed stellar polygon made by connecting next k points. This highlights that students should reflect on what they have done in task 2 (reflective stage) and do more activities for integrating them (analytic stage) according to the learning stages in the "two-axis process model" of mathematical understanding before proceeding to task 3 which was of a higher level.

Now based on the above suggestions, we can make an alternative plan for the 8th grade lesson on "regular stellar polygon" in a lower secondary school mathematics classroom which is as follows. When a teacher intends to help his/her students develop their mathematical understanding according to three levels of mathematical entities, relation of the entities, and general relation in a vertical axis of the "two-axis process model", it is suggested to set up three learning stages in a horizontal axis of the model by posing the following alternative tasks such as task 1^+ in a regular nine-pointed stellar polygon, task 2^+ in a regular nine-pointed stellar polygon, and task 3^+ in a regular 12-pointed stellar polygon before posing the generalized task 4^+ in a regular n-pointed stellar for asking students to formulate the measure of an angle in a general way.

Task 1^+: "Draw a regular nine-pointed stellar polygon made by connecting next k points for $k = 1,2,3,4$".

Task 2^+: "Find out and make a table of the measures of tip angle in a regular nine-pointed stellar polygon made by connecting next k points for $k = 1,2,3,4$".

Task 3^+: "Find out and make a table of the measures of tip angle in a regular 12-pointed stellar polygon made by connecting next k points for $k = 1,2,3,4,5,6$".

Task 4^+: "Represent the measure of a tip angle in a regular n-pointed stellar polygon made by connecting next k points as an algebraic expression with letters n and k".

The lesson plan for the above tasks needs two fifty-minute sessions because it is very difficult for 8th graders in one session to investigate and represent algebraically the general relation among the measures of a tip angle in a regular n-pointed stellar polygon made by connecting next k points. If t is the measure of a tip angle, $n \geq 3$, and $1 \leq k \leq n-1$, then

$$t = \left|180° - \frac{360°}{n} \times k\right|.$$

In the first session, in order to help the students improve their mathematical understanding from the level of mathematical entities (the measures of tip angle in a regular nine-pointed stellar polygon made by connecting next k points) to the level of the relation of them (the relation among the measures of tip angle in a regular nine-pointed stellar polygon), task 1^+ is posed for the intuitive stage and task 2^+ is posed for the reflective and analytic stages. Then in the second session, task 3^+ is posed for the intuitive stage and task 4^+ is posed for the reflective and analytic stages in order to encourage the students to generalize the relation and help them improve their mathematical understanding from the level of the relation among the measures of tip angles in a regular nine-pointed stellar polygon to the level of the general relation (the general relation among the measures of a tip angle in a regular n-pointed stellar polygon made by connecting next k points).

The mathematical topic of "regular stellar polygon" is very rich in improving the students' mathematical understanding in a lower secondary school classroom. We have to examine the effectiveness of the alternative lesson plan for helping the students improve their mathematical understanding up to be an expected higher level in a classroom.

Notes

This article is a modified English version of Koyama (2006). It is partly based on the author's work supported by a grant-in-aid for scientific research from the Japan Society for the Promotion of Science (JSPS), grant No.16530591 and No.22530978. Any opinions, findings and conclusions or recommendations are those of the author and do not necessarily reflect the views of the JSPS.

Acknowledgement

The author deeply thanks a mathematics teacher Mr. Tominaga and his all anonymous students whose work has provided the content for this chapter.

References

Byers, V., & Herscovics, N. (1977). Understanding school mathematics. *Mathematics Teaching, 81*, 24-27.

Herscovics, N., & Bergeron, J.C. (1988). An extended model of understanding. *Proceedings of the 10th PME-NA*, 15-22.

Koyama, M. (1992). Models of understanding in mathematics education. In the memorial publishing board for the retirement of professor Kazuo Iwago (Ed.), *New developments in mathematics education* (pp. 172-184). Tokyo, Japan: Seibunsha. (In Japanese)

Koyama, M. (1993). Building a two-axis process model of understanding mathematics. *Hiroshima Journal of Mathematics Education, 1*, 63-73.

Koyama, M. (1995a). On reflective thinking. In T. Nakahara (Ed.), *Report of practical research on designing lesson based on the constructive approach in mathematics education* (pp. 30-38), Hiroshima, Japan: Hiroshima University. (In Japanese)

Koyama, M. (1995b). Characterizing eight modes of the transcendent recursive model of understanding mathematics. *Journal of Japan Academic Society of Mathematics Education: Research in Mathematics Education, 1*, 19-28. (In Japanese)

Koyama, M. (1997a). Research on the complementarity of intuition and logical thinking in the process of understanding mathematics. *Hiroshima Journal of Mathematics Education, 5*, 21-33.

Koyama, M. (1997b). Mathematics learning and process of understanding. In Japan Society of Mathematical Education (Ed.), *Rethinking lesson organization in school mathematics* (pp. 135-149). Tokyo, Japan: Sangyotosho. (In Japanese)

Koyama, M. (2005). The principles and methods for designing mathematics lesson based on the "two-axis process model". *Journal of Japan Curriculum Research and*

Development Association: The Bulletin of Japanese Curriculum Research and Development, 28(4), 61-70. (In Japanese)

Koyama, M. (2006). Research on the process of understanding in learning mathematics (I): Analysis of the lesson on "stellar polygon" at second grade in lower secondary school. *Journal of Japan Academic Society of Mathematics Education: Research in Mathematics Education, 12*, 71-81. (In Japanese)

Koyama, M. (2007). Research on the process model of mathematical understanding in primary mathematics education. *Journal of Japan Society of Mathematical Education: Reports of Mathematical Education, 88*, 25-35. (In Japanese)

Koyama, M. (2010). *Research on the process model of mathematical understanding in primary mathematics education*. Tokyo, Japan: Seibunshinsha. (In Japanese)

Ministry of Education, Science, Sports, Culture and Technology. (1999). *Guidebook for the lower secondary school mathematics in the Course of Study (1998)*. Osaka: Osakashoseki. (In Japanese)

Nakahara, T. (1995). *Research on the constructive approach in mathematics education*. Tokyo, Japan: Seibunsha. (In Japanese)

Pirie, S., & Kieren, T. (1989). A recursive theory of mathematical understanding. *For the Learning of Mathematics, 9*(3), 7-11.

Pirie, S., & Kieren, T. (1994). Growth in mathematical understanding: How can we characterise it and how can we represent it?. *Educational Studies in Mathematics, 26*, 165-190.

Skemp, R.R. (1982). Symbolic understanding. *Mathematics Teaching, 99*, 59-61.

Tominaga, K. (2003). *A mathematics lesson plan for lower secondary school*. Paper presented at the open-lesson study in a lower secondary school attached to Hiroshima University, Hiroshima, Japan. (In Japanese)

van Hiele, P.M., & van Hiele-Geldof, D. (1958). A method of initiation into geometry at secondary schools. In H. Freudenthal, (Ed.), *Report on method of initiation into geometry* (pp. 67-80). Groningen Wolters.

Wittmann, E. (1981). The complementary roles of intuitive and reflective thinking in mathematics teaching. *Educational Studies in Mathematics, 12*, 389-397.

Chapter 15

Mathematics Learning Episode that Promotes Reflective Thinking Among Elementary Pupils

Auxencia Alarcon LIMJAP

A framework of learning and teaching referred to as the Learner Centered Learning Environment (LCLE) guides the teachers of La Salle Greenhills, a sectarian school for boys in the Philippines, in the preparation of modules for classroom teaching. With the goal to promote reflective thinking among students, teachers employ learning episodes that elicit students' prior knowledge and lead them to reflect and inquire about the new topic. This chapter describes how a teacher nurtures reflective learners in a Grade 4 mathematics class. With the LCLE framework as a guide, the learning episode shows how the teacher generates responses from pupils using prompts and questions that effectively engages elementary pupils in reflective learning.

1 Introduction

Reform movements in education worldwide include the institutionalization of constructivist inspired pedagogical frameworks for teaching and learning in the Philippines both at the national level through the Department of Education (2012) and at the school level. La Salle Greenhills (LSGH) is a sectarian school for boys in Metro Manila that anchored its Learner Centered Learning Environment (LCLE) framework on the constructivist philosophy and the 14 Learner Centered

Psychological Principles of the American Psychological Association (APA) in 1997. These 14 principles are divided into factors namely 1) cognitive and metacognitive, 2) motivational and affective 3) developmental and social and 4) individual difference factors influencing learners and learning. Aware of the importance of mathematics in the school curriculum, teachers are encouraged to take a learner centered perspective in teaching it.

A thorough rethinking of the theories of learning still radically affects school mathematics education. Learner centered principles support the conviction that mathematics learning should consist of concept development and deep understanding rather than behavior or skills. National Research Council (2002, pp. 9-16) identifies the five strands of mathematical proficiency to be as follows: a) understanding the concepts b) computing fluently c) applying concepts to solve problems d) reasoning logically and e) engaging with mathematics. To achieve meaningful learning, students should learn mathematics with understanding. National Research Council (2002, p. 10) further claims that "knowledge learned with understanding provides a foundation for remembering or reconstructing mathematical facts and methods, for solving new and unfamiliar problems, and for generating new knowledge."

This perspective supports the constructivist view of learning as an active process. Janvier (1996, p. 452) commented that "learning will always be achieved in a constructivist manner whatever you do...the subject cannot do otherwise but learn constructively." According to Noddings (1990), "mathematical knowledge is constructed at least in part, through a process of reflective abstraction." Herscovics (1996, p. 359) describes reflective abstraction as consisting of "separating concept from the procedure, [and then] generalizing the concept" through some complex mathematical processes. Learners "create (construct) new mathematical knowledge by reflecting on their physical and mental actions" (Reys, *et al.*, 2004, p. 21).

Reflective thinking comes from the work of John Dewey in his book How to Think in 1933. Dewey views reflection as a meaning making process that moves a learner through various experiences with deeper understanding of the relationships. It is a systematic, rigorous, disciplined way of thinking that cultivates significant learning. Thinking reflectively happens in a community of learners as they interact with one another. It requires attitudes that value one's personal and intellectual growth and those of others. It involves "personal commitment to the idea that mathematics makes sense and that given reasonable effort, they can learn it and use it, both in school and outside school" (National Research Council, 2002, p. 16).

More recent contributions on reflective practice to educational theory is attributed to the research inquiry of Jack Mezirow in his transformational learning theory. Mezirow's theory otherwise called transformative learning in the classroom is characterized by a student's life experiences, critical reflection, and rational discourse. He emphasizes the need to develop in students the ability to communicate their ideas for them to be able to engage in critical discourse (Mezirow, 1991). According to Phan, Mezirow advocates four stages of reflective thinking namely

1) Habitual action is a mechanical and automatic activity that is performed with little conscious thought.
2) Understanding is learning and reading without relating to other situations
3) Reflection concerns active, persistent and careful considerations of any assumptions or beliefs grounded in our consciousness
4) Critical reflection is considered as a higher level of reflective thinking that involves us becoming more aware of why we perceive things, the way we feel, act and do (Mezirow, 1991 & 1998, as cited by Phan, 2006, p. 583)

This study shows how a mathematics teacher implements learner centered principles of the LCLE framework to nurture reflective learners in her class. The different phases of the framework are presented in the next section.

2 Learner Centered Learning Environment

The Learner Centered Learning Environment Handbook of LSGH by Rapatan (2004, p. 7) divides the Grade School LCLE into four main parts or phases of the instructional system. These are

I. Identifying the Students' Prior Knowledge
II. Providing Engaging Interactions
III. Coaching Students' Transformation and Mastery
IV. Evaluating and Verifying Students' Performance.

The LCLE framework provides teachers with a perspective, specific strategies and techniques for a dynamic and educationally enriching learning experiences for the students. See appendix A for the diagrammatic illustration of the expected teacher actions in this framework. The phases are as follows.

2.1 *Identifying the students' prior knowledge*

This phase gives teachers the opportunity to check the students' level of conceptual understanding, their procedural knowledge and prerequisite skills needed for the new topic. According to Reys *et al.* (2004, p. 25) "the challenge is not only to develop these types of knowledge but also to understand relations between them." Teachers should create situations that build connections from students' past lessons and experiences and generate their understanding of concepts and their relationships.

Problem situations that are familiar to the students may help elicit their prior knowledge. Furthermore, these situations can "build on what students know, mathematically and experientially, including creating and connecting [them] with stories that both contextualise and establish a rationale for the learning" (Sullivan, 2011, p. 28).

Misconceptions of students usually surface at this phase. Teachers should take this opportunity to engage students in dialogues or give activities which require skills needed to tackle the new topic. Kaur and Toh (2012, p. 3) noted that "[c]entral to all mathematics lessons are mathematical tasks." Even at the elementary level "students are never too young or the mathematics content too basic for students to engage in meaningful reasoning and communications during mathematics lessons" (Kaur, 2012, p. 86). "[T]he tasks with which students become engaged determine[d] not only what content they learn but also how they can engage in the academic work of the classroom and therefore learn how to think about, develop, make sense of and apply the content knowledge they encounter in the task" (Rahim, Hogan, & Chan, 2012, p. 13). Teachers should start asking open ended questions where students can elaborate their understanding of the topic. They should process the responses of students and ask them to synthesize or summarize their answers. Students should be able to explain how they get their answers.

This phase is where both teachers and students should engage in reflective thinking. Teachers should reflect on how best to elicit the varied conceptions of students in a creative way. With the guidance of the teacher, students should be able to reflect on their understanding and past learning and use them to undertake the initial tasks for the new lesson. This is the PK phase, referring to prior knowledge.

2.2 *Providing engaging interactions*

This phase provides students with more opportunities to deal with the new concepts by working with activities that are authentic in context. Since students construct their knowledge by building on their experiences, then the teacher should "[e]ngage students by utilising a variety of rich and challenging tasks that allow [them] time and opportunities to make decisions, and which use a variety of forms of representation" (Sullivan, 2011, p. 28). The resources needed to fulfill the task should be available including technology, media and reference materials.

It is possible to stimulate students to engage in creating knowledge if the tasks provide students with "opportunities to make decisions about either the strategy for solving the task or the process they will adopt for addressing the task goal...provide some degree of challenge, address important mathematical ideas and foster communication and reasoning" (Sullivan, 2011, p. 31). Students can do this individually or in groups.

Swan (2005 as cited by Sullivan, 2011, p. 33) encourages teachers to "use rich collaborative tasks ... that emphasise methods rather than answers, facilitate connections between topics, support cooperative group work, build on what the students bring to the session, and explore common misconceptions." Mathematics educators believe that "[p]roficiency is much more likely to develop when mathematics classroom is a community of learners rather than a collection of isolated individuals" (National Research Council, 2002, p. 26). This phase supports the learner centered principles of social constructivism.

The teacher should be able to monitor collaborative tasks with proper orientation on the conduct of the group work and efficient monitoring of the group's progress. Students should be encouraged to "generate and share solution methods, [where] mistakes are valued as opportunities for everyone to learn, and correctness is determined by the logic and structure of the problem, rather than by the teacher" (National Research Council, 2002, p. 26). This is the ET phase referring to engaging task.

2.3 *Coaching students' transformation and mastery*

This phase is very crucial because this is the point where students experience cognitive conflicts or dissatisfaction with their current understanding of the concept. The role of the teacher as coach is very important. As a facilitator of learning, the teacher should provide the scaffolding or guide to help students focus their inquiry to the path that leads to the correct concept. The scaffold can come in various forms like leading probing questions, feedback to the initial findings of students, or concept maps which help students visualize the core concepts and the supporting details (Rapatan, 2004, p. 18).The teacher should also help students recognize the intelligibility, plausibility and fruitfulness of the new conception.

The role of the mathematical community consisting of other learners and teacher is "to provide the setting, pose the challenges, and offer the support that will encourage mathematical construction" (Davis, Maher, & Noddings, 1990, p. 3). In this setting, students will exert effort to explore and find the correct meaning of the concepts.

Interaction of students with one another is very crucial too. One way of ensuring fruitful interaction is by providing students with task sheets or guide sheets in their exploration. As soon as students recognize the need to explore new ideas and change their perspective or viewpoint, transformation begins. The teacher should help students not only assimilate but accommodate the new concept as well. According to Steffe and Wiegel (1996, p. 491) "[a]ccommodation...accounts for qualitative changes in mental or physical actions, operations, images and schemes." They add that Janvier's position that any learning is bound to be constructivist "fits our conception of learning as accommodation" (p. 491). Once students accommodate the new conception, then learning becomes meaningful to them. They achieve deep comprehension, recognize relationships, and elaborate connections. They are able to transfer their learning to different situations and gain mastery of the concept.

To facilitate meaningful understanding, the teacher can ask process questions such as "What rule can we form which puts together for us these new ideas? How can we tell if our actions and decisions are right? Which situations led you to the generalization that you made? How did you put together those ideas? What materials or references helped you most in understanding the topic? What help did these give you in forming your rule or generalization? What other situations can we use to apply the rule or generalization? What task did you find most difficult to do?" (Rapatan, 2004, pp. 20-21). This is the TM phase, referring to transformation and mastery.

2.4 *Evaluating and verifying students' performance*

There are certain metacognitive skills that can be developed for meaningful and significant learning. One such skill is the ability to regulate one's own learning. This is achieved if the students are given the opportunity to assess their performance and that of their peers. If there is a conscious effort by students to monitor their own progress, and think about their mathematical thinking, then they may gain mastery of the lesson and improve their performance. According to Reys *et al.*

(2004, p. 30) teachers can ask students such questions as "What errors do you make most often in mathematics? Why do you think you make them?"

Open ended prompts can assess how meaningful students' learning has been. Engaging students in such authentic assessments as performance assessment reveals students' level of conceptual understanding and ability to work independently. Reflection journals can determine the extent of the cognitive change of students. The teacher can ask process questions like "Discuss briefly how you developed your concept. How did your ideas change from the start of the lesson to the end of the activities? What differences do you see? How will you rate your understanding from a scale of 1 to 10 with 10 representing clear understanding? Why? How can you raise your understanding if your score is less than 10?" (Rapatan, 2004, p. 24). This is the EV phase referring to evaluation and verification.

3 LCLE Learning Episode on Fractions

The teacher of the Grade 4 pupils, Teacher Cristine, was interviewed before she taught the topic on subtracting mixed numbers from whole numbers. This interview revealed the alignment of her beliefs about teaching and learning with the LCLE principles and how the principles were applied in her lesson plan.

According to Teacher Cristine, she applies the constructivist philosophy whenever she writes lesson plans. She makes the boys understand how the procedures come about and the importance of knowing them. She asks a lot of questions from the start of the lesson to engage the pupils in the learning process. She finds out what her pupils already know by giving them brainteasers or games that are interesting to them. She also conducts reviews. She admits that her pupils just wait for the mathematical procedures and care less about the reasons behind the procedures. But she uses visual representations and requires them to provide their solutions with visual representations as well to develop their conceptual understanding.

The different phases of Teacher Cristine's lesson about subtracting mixed numbers from whole numbers are presented in the following sections.

3.1 *Eliciting students' understanding of subtraction of fractions*

The first phase dealt with a review on subtracting fractions. However, the process took long as the teacher checked on the students' conceptual understanding of subtracting fractions through the inquiry based approach. She asked questions which Reys *et al.* (2004, p. 28) described as having the potential to "encourage critical thinking, establish relationships, and promote meaningful connections." She also used appropriate visual representations in accordance with Sullivan's (2012) principle of "differentiating challenges" and Reys *et al.* (2004) recommendation to "use manipulatives to aid learning."

As the teacher asked questions on subtracting fractions, the pupils found it difficult to justify the procedures. But with the teacher's art of asking process questions, she was able to draw desired generalizations from the pupils as follows:

[PK55] T: What are the steps in subtracting similar fractions? Enumerate all the steps.

[PK56] S1: Subtract the numerator, copy the denominator and then reduce the answer to lowest terms.

[PK57] T: There's a reason for copying the denominator. Why unlike whole numbers why don't they subtract the denominator? Why say copy the denominator? Why do we copy the denominator?

[PK58] S1: Because we don't add or multiply the denominators.

[PK59] T: Yes we don't add because operation is subtraction right? That is why we don't add right? And we don't add and we don't multiply the denominators.

[PK60] T: Alright! Why can't we subtract the denominator? Why is the rule telling us to copy instead.

[PK61] S2: The denominator does not get smaller if we subtract fractions. It still remains the same.

As pointed out by Herscovics (1996), this is an indicator that the students reflected on the given situation and arrived at a synthesis or a generalization. The teacher made more attempts to help the pupils understand the underlying concept with the help of visual representations. She gave a story problem and showed an illustration of chocolate bars on the television screen. She provided scaffolds referred to by Sullivan (2012) as "enabling prompts".

Teacher Cristine asked a student to read the problem.

[PK77] S3: I have ¾ of the chocolate bar. My sister has ¼ of the chocolate bar. What part of the fraction do I have more?

[PK78] T: How much more (chocolate) do I have? So we have same chocolates but what is being asked?

[PK79] T: How much more do I have visually? How much more do I have in the first figure?

[PK80] T: As you can see we have the same (chocolate bars) but I just have $^{2}/_{4}$ more than my sister? So that is 1 and 2 more. (Points to the 2 parts)How many parts in all do I have in the first figure?

[PK81] S4: 4

[PK82] T: Did it increase? Did it change? Do the chocolates still have small parts?

[PK83] C: Yes

[PK84] T: Yes of course, because if I subtract the fourths and fourths what would be the denominator? If I subtract the denominators it would be?

[PK85] C: 0

[PK86] T: So there will be no chocolates. So that's the reason why we copy [the denominator].

At this point, students' responses to the questions posed in class, were indicators of Mezirow's first two levels of reflection called habitual action and understanding. As the teacher posed more guided questions, the pupils achieved higher levels of reflection such as in the following dialogue.

[PK118] T: Why do we really need to give the answer in lowest term? What is the rule? Why do you think?
[PK119] S5: To make the number smaller.
[PK120] T: To make the number smaller but they are equal!
[PK121] S6: To make it easier to read.
[PK122] T: To make it easier to read. So it is being...
[PK123] S6: Simplified.
[PK124] T: To make the number smaller, but they are equal!
[PK 125] S6: To make it easier to read.

The teacher gave the example $^{6}/_{12}$ and $^{5}/_{10}$ which are equal to the same fraction ½ and pointed out how easy it is to visualize the simplest form, ½. The class agreed with S6 that "the lowest terms of the fraction is equal to the original fraction and makes the original fraction easier to read." It is the teacher's quick feedback that corrected the claim of S5 that "the lowest terms of a fraction make the original fraction smaller." On the other hand, S6 manifested reflective thinking when he stated his conclusion.

While this phase focused on the prerequisite skills of pupils, the teacher did not dwell on procedures alone. She raised pupils' level of reflection from the habitual action that is algorithmic or procedural to an understanding of the relationships of the concepts. As pupils made generalizations, they achieved the third level of reflection defined by Mezirow (1991).

3.2 *Providing tasks on subtracting mixed number from whole number*

The teacher asked more guiding questions and used visual representations in this phase. The pupils used a circle to visualize $1 - {}^{2}/_{3}$. When the teacher asked 'What did I do with 1 whole?' the class (C) answered 'it was made into a fraction.' Then they said that 'it was written as $^{3}/_{3}$.' However, this phase also reveals the inclination of the pupils to be mechanical and procedural without giving the process of subtraction much thought. This is the case when they were asked to represent $8 - 5\ {}^{1}/_{3}$ visually. Some pupils immediately answered 3 when asked to find the difference, which is a manifestation of the habitual action of subtracting 8 and 5. However, the teacher skillfully led the pupils to understand the relationships involved in the process using visualization through a series of well thought of questions. She repeatedly asked the following questions:

[ET191]	T:	How many circles representing the whole number will I draw here?
[ET202]	S9:	8
[ET203]	T:	Eight, yes that is what I'm asking 1, 2, 3, 4, 5, 6, 7, 8. How many whole numbers do I deduct here?
[ET208]	S10:	5
[ET211]	T:	Alright, is this the final answer?
[ET212]	C:	No
[ET213]	T:	Why what did I not subtract?
[ET217]	T:	I did not subtract the 3 whole numbers to what?
[ET218]	C:	Fraction
[ET219]	T:	What fraction?
[ET220]	C:	$^{1}/_{3}$
[ET221]	T:	How many parts will I cut this into? Everybody
[ET220]	C:	3
[ET221]	T:	Alright because that's 3 parts, how many will I take away? Everybody

At this point, the class gave different answers. When the student who was called gave a wrong answer, some pupils howled in protest. This may seem unruly, but this is an indication that the pupils were listening and were engaging in critical reflection while the teacher asked questions. They were able to give the correct answer 2 $^{2}/_{3}$.

3.3 *Assessing learning to coach students' transformation and mastery*

Teacher Cristine integrated the TM and EV phases in an activity with guided prompts. She used this activity to assess how much her pupils learned from the discussion. They were told to work in pairs. As she moved around to facilitate the work of the dyads, she coached the students and verified how well they understood the new topic. She challenged the pupils' understanding of the lesson and encouraged them to reflect and think critically. The next sections show this.

4 Reflective Thinking in a Procedural Task

The key actions expected from students in a LCLE class are reflecting, interacting, transforming and evaluating as they work collaboratively using strategic reasoning and executive control (Rapatan, 2004). See appendix B for the diagrammatic illustration of expected student actions in the framework. They should be able to use a variety of strategies to help them acquire the baseline knowledge and information, make meaning of this knowledge, and transfer what they learned to new situations. Even in tasks that are procedural in nature, their strategic skill can still be used as they come up with a plan or a schema. Watson and Sullivan (2008, as cited by Sullivan, 2011, p. 7) describe strategic competence as "the ability to formulate, represent and solve mathematical problems." As students implement the plan or strategy to use mathematics, they exhibit their understanding of the new knowledge and their exercise of executive control (Rapatan, 2004, p. 11).

4.1 ***Representations and reflections in dyads***

A seatwork in dyads was the culminating activity of this learning episode on subtracting fractions. Pupils were asked to respond to the following questions found in Figure 1.

Subtract and draw the visual representation of the answer.

1. $1 - {}^{5}/_{6}$
2. $8 - 3\,{}^{2}/_{3}$
3. In question # 1
   A. What did you do to 1 whole before you subtracted?
   B. Why did you choose that fraction?
4. In question # 2
   C. What did you do to 8?
   D. Do you subtract 8 and 3?
   E. How did you answer # 2?

Figure 1. Seatwork in dyads on subtraction of fractions

The level of pupils' understanding was assessed using their responses. Stages of reflective thinking posited by Mezirow (1991) were used to describe their mathematical thinking. Figure 2 shows the work of Seth and Jose.

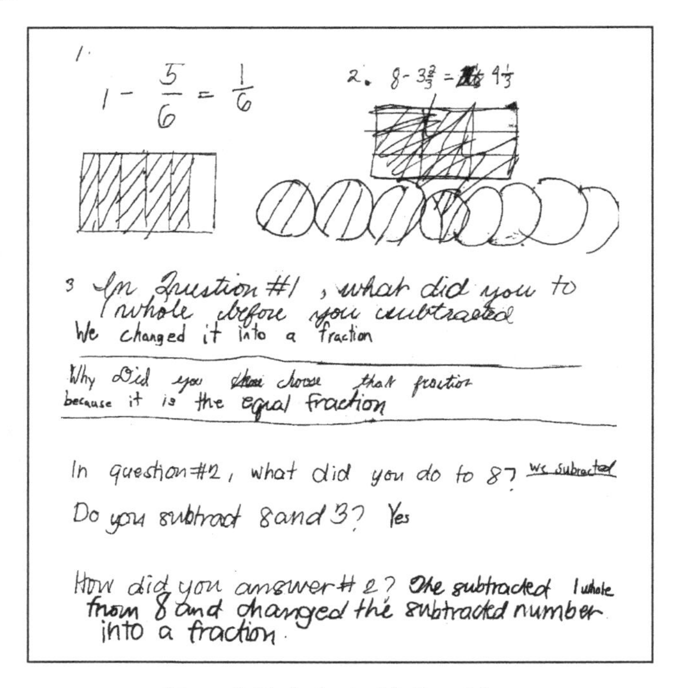

Figure 2. Task sheet of Seth and Jose

Seth and Jose negotiated their ideas as they wrote their answers. The erasures were made as they decided on the best model. They solely relied on the illustrations for their answers to the computations. The model provided them with the mathematical context for subtracting a fraction from a whole number. While they were able to give the correct answer, it seems that they were not able to make meaningful connections. Their answers to question numbers 3A-E are not linked to their solutions in items 1 and 2. They wrote that they 'changed [1] into a fraction,' which is not found in their solution to question number 1. They answered the next process question without doing the necessary procedure. Their answers to the last three questions 3C, D, and E are also vague. Their responses to the process questions may be indicators of Mezirow's first two stages of reflection. While there is conscious thought involved with Seth and Jose's use of the visual strategy, leading to correct answers, the meaningful bridges from the model to the mathematics are not yet there. Reys *et al.* (2004) admit that establishing these links and connections between the model and mathematics is really a challenge (p. 28).

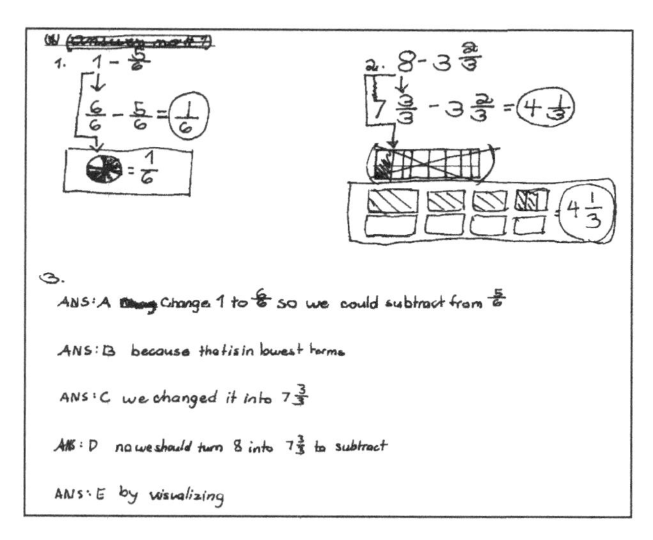

Figure 3. Task sheet of Mike and Dos

Figure 3 shows the work of Mike and Dos. Mezirow's stage 3 of reflection is apparent when Mike and Dos described clearly in items 3A, C and D the solution that they wrote in items 1 and 2. But the answer to 3B is vague, and the answer to 3E is very general. However, a good amount of reflection as described in Mezirow's stage 3 is evident in the work of Mark and Dos.

The work of Andrew and Tam, shown in Figures 4 and 5, is a perfect example of the attainment of Mezirow's critical reflection. The accuracy and clarity of their responses exhibit both their procedural and conceptual understanding. All indications of complete understanding are in the task sheet of Andrew and Tam. The answers are interrelated. They explicitly wrote in 3B that the fractional form of the whole number should have 'the same denominator as the fraction in the subtrahend.'

They explained the procedure clearly in 3E. Thus in terms of Mezirow's stages, they are at the fourth stage of critical reflection.

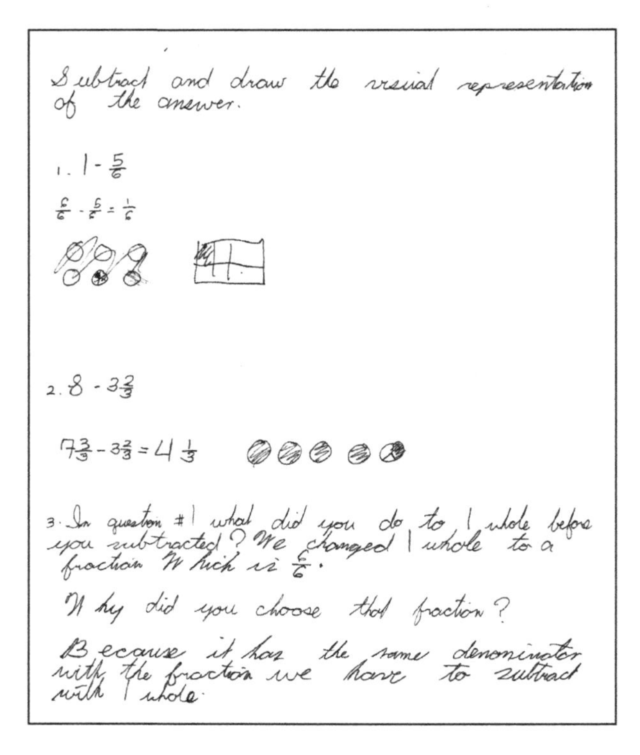

Subtract and draw the visual representation of the answer.

1. $1 - \frac{5}{6}$

$\frac{6}{6} - \frac{5}{6} = \frac{1}{6}$

2. $8 - 3\frac{2}{3}$

$7\frac{3}{3} - 3\frac{2}{3} = 4\frac{1}{3}$

3. In question #1 what did you do to 1 whole before you subtracted? We changed 1 whole to a fraction Which is $\frac{6}{6}$.

Why did you choose that fraction?

Because it has the same denominator with the fraction we have to subtract with 1 whole.

Figure 4. Front side of the task sheet of Andrew and Tam

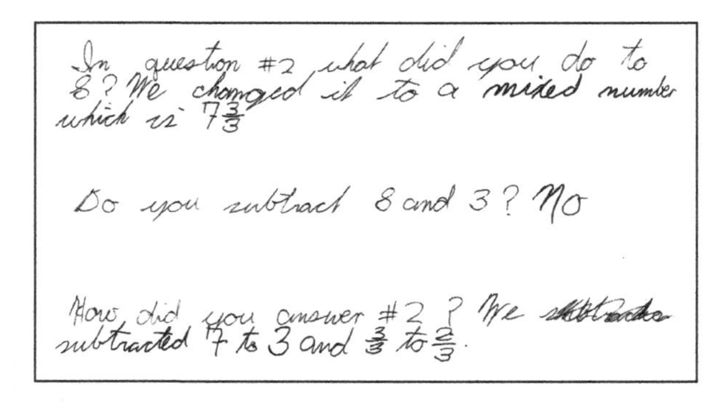

In question #2 what did you do to 8? We changed it to a mixed number which is $7\frac{3}{3}$

Do you subtract 8 and 3? No

How did you answer #2? We subtracted 7 to 3 and $\frac{3}{3}$ to $\frac{2}{3}$.

Figure 5. Back side of the task sheet of Andrew and Tam

4.2 *Manifestations of reflective thinking*

The Grade 4 pupils participated meaningfully in the class discussion. They recited when called individually, responded collectively to the prompt questions, reacted to the wrong answers of their classmates, and asked questions during the seatwork. The boys listened attentively while their teacher was explaining.

After the class, selected pupils were interviewed to find out how well they could reflect on 'what they know and still need to know'. They were asked to think about their solutions in the task sheet. All students who were interviewed were able to explain their responses to the task sheet. A good example is the explanation of Matthew when asked about his solutions and answers to the items 1) $1 - {}^{5}/_{6}$ and 2) $8 - 3\ {}^{2}/_{3}$.

Interviewer: What made you answer ${}^{1}/_{6}$?
(refers to the solution for $1 - {}^{5}/_{6}$)

Matthew: Because it is a whole number and you are turning it into a fraction if you are subtracting a fraction, then you turn the whole number to ${}^{6}/_{6}$ since the denominator here (points to ${}^{5}/_{6}$) is 6. So I got ${}^{1}/_{6}$.

Interviewer: What about this one. Can you explain to me what you did here? (points to the solution for $8 - 3\ {}^{2}/_{3}$).

Matthew: This time we are dealing with a mixed number. Its not just a fraction. We're dealing with a mixed number. So you like ah...if here (points to $1 - {}^{5}/_{6}$) you turn thiswhole into a fraction (points to 1), of course here (points to $8 - 3\ {}^{2}/_{3}$), you need to turn...its ...okay...this is a fraction (points again to $1 - {}^{5}/_{6}$), you need to turn this into a fraction (points again to 1). So this one (points to $8 - 3\ {}^{2}/_{3}$), since this is a mixed number, (points to $3\ {}^{2}/_{3}$), you kinda need to turn this (points to 8) into a mixed number. So same thing.

In his paper he wrote 'Turn 8 into 7 and ${}^{3}/_{3}$ so that you may have something to subtract for the whole number and fraction. Subtract 3 from 7 and ${}^{2}/_{3}$ from ${}^{3}/_{3}$.' The pauses in Matthew's answer reveal his conscious effort to reflect on the mathematical procedure and the underlying concept.

5 Conclusion

This chapter shows how a teacher applies her pedagogical content knowledge to promote reflective thinking among her pupils. Her beliefs about teaching and learning are in consonance with the LCLE framework of La Salle Greenhills. By adhering to the four phases of LCLE, she consciously applies constructivist based principles of teaching and learning mathematics to ensure that her pupils learn meaningfully. By using an inquiry based teaching focused on the thinking skills and processes, she creates an environment for learning that generates reflection among elementary pupils. The responses of the pupils to the prompt questions at different phases of LCLE manifest different stages of reflection. Evidences show that her pupils gained understanding of subtracting fractions by connecting visual images with the mathematical concept and by engaging in reflection.

Acknowledgements

I am grateful to La Salle Greenhills Grade School's principal, Mr. Jose Ramelle Javier, Math Dept. chair, Mr. Noel Menor, and Grade 4 teacher Ms. Cristine Ann Mallari, for allowing me and my assistant Mr. Geoffrey Reuel Pasague to conduct this research; to Professor Berinderjeet Kaur for this opportunity to contribute to this yearbook and for generously accommodating me in Singapore during the AME-SMS 2012 Conference; and to Professor Judy Anderson for giving valuable suggestions for the improvement of this chapter.

References

Davis, R., Maher, C., & Noddings, N. (1990). Introduction: Constructivist views on the teaching and learning of mathematics. In R.B. Davis, C.A. Maher, & N. Noddings (Eds.). *Journal for research in mathematics education* monograph (pp. 1-3) VA: NCTM.

Department of Education. (2012). *K to 12 Curriculum Guide Mathematics* (Grade 1 to Grade 10). Pasig City: Republic of the Philippines, Department of Education.

Dewey, J. (1933). *How we think*. D.C. Heath & Co.

Herscovics, N. (1996). The construction of conceptual schemes in mathematics. In L.P. Steffe & P. Nesher (Eds.). *Theories of mathematical learning*. (pp. 351-379). New Jersey: Lawrence Erlbaum Associates.

Janvier, C. (1996). Constructivism and its consequences for training teachers. In L.P. Steffe & P. Nesher (Eds.). *Theories of mathematical learning*. (pp. 449-462). New Jersey: Lawrence Erlbaum Associates.

Kaur, B. (2012). Some "what" strategies that advance reasoning and communication in primary mathematics classrooms. In B. Kaur & T.L. Toh (Eds). *Reasoning, communication and connections in mathematics* (pp. 75-88). Singapore: World Scientific.

Kaur, B., & Toh, T.L. (2012). Reasoning, communication and connections in mathematics: An introduction. In B. Kaur & T.L. Toh (Eds). *Reasoning, communication and connections in mathematics* (pp. 1-10). Singapore: World Scientific.

Learner-Centered Principles Work Group of the American Psychological Association's Board of Education Affairs. (1997). *Learner-centered psychological principles: A framework for school reform and redesign*. Retrieved on October 2, 2012 from http://www.apa.org/ed/governance/bea/learner-centered.pdf

Mezirow, J. (1991). *Transformative dimensions of adult learning*. San Francisco, CA: Jossey-Bass.

National Research Council (2002). *Helping children learn mathematics*. Washington, DC: National Academy Press.

Noddings, N. (1990). Constructivism in mathematics education. In R.B. Davis, C.A. Maher, & N. Noddings (Eds.). *Journal for research in mathematics education* monograph (pp. 7-18) Virginia: NCTM.

Phan, H.P. (2006). Examination of student learning approaches, reflective thinking, and epistemological beliefs: a latent variables approach. *Electronic Journal of Research in Educational Psychology*, *10*(4[3]), pp. 577-610.

Rahim, R.A., Hogan, D., & Chan, M. (2012). The epistemic framing of mathematical tasks in secondary three mathematics lessons in Singapore. In B. Kaur & T.L. Toh (Eds). *Reasoning, communication and connections in mathematics Yearbook 2012 Association of Mathematics Educators* (pp. 11-56). Singapore: World Scientific.

Rapatan, M.Q. (2004). *Learner-centered learning environment handbook for LSGH*. Mandaluyong: La Salle Greenhills.Bell

Reys, R., Lindquist, M., Lambdin, D., Smith, N., Suydam, M., Niess, M., Erickson, D., & Higgins, K. (2004). *Helping children learn mathematics*. 7th ed. New Jersey: John Wiley & Sons, Inc.

Steffe, L., & Wiegel, H. (1996). On the nature of a model of mathematical learning. In L. Steffe, P. Nesher, P. Cobb, G. Goldin, & B. Greer (Eds.). *Theories of mathematical learning* (pp. 477-498). New Jersey: Lawrence Erlbaum Associates.

Sullivan, P. (2011). *Teaching mathematics: Using research-informed strategies.* Victoria: Australian Council for Educational Research. Retrieved October 18, 2012 from http://research.acer.edu.au/aer/13

Appendix A (Rapatan, 2004, p. 11)

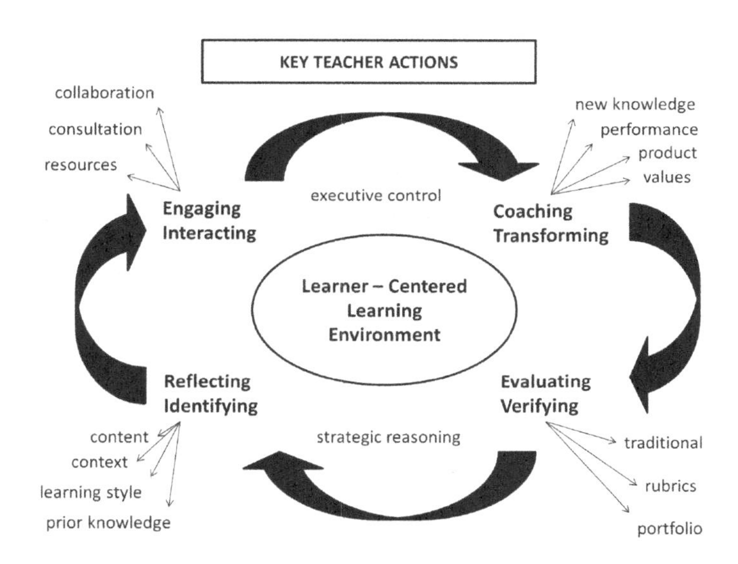

Appendix B (Rapatan, 2004, p. 12)

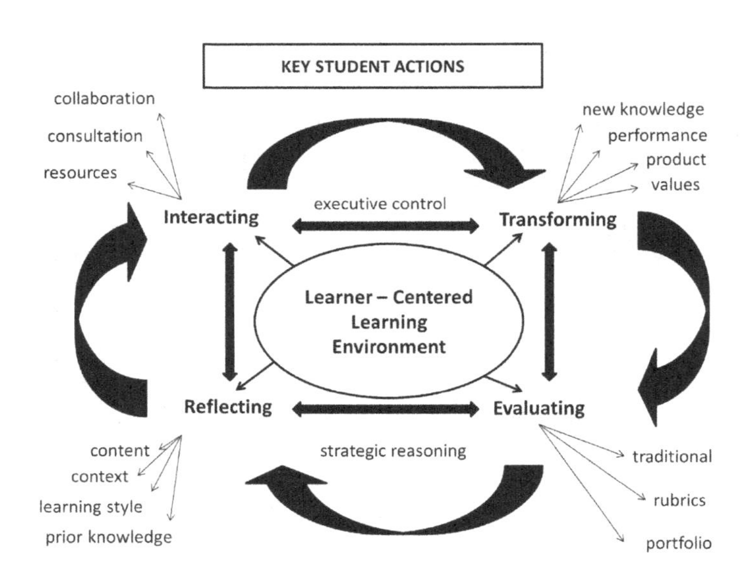

Contributing Authors

Kim BESWICK taught mathematics and science in a number of secondary schools for 13 years before joining the University of Tasmania in 2000 where she is currently an Associate Professor in mathematics education and Associate Dean (Research) for the Faculty of Education. Kim is involved in teaching primary mathematics curriculum at undergraduate level and both mathematics curriculum and research methods for postgraduate students. Her research interests include beliefs and knowledge that underpin the practice of mathematics teachers, professional learning of mathematics teachers/educators, equity in education, and the development of students' understanding of concepts related to proportional reasoning. Kim supervises research higher degree students in these and other areas of mathematics education and has published more than 60 peer reviewed book chapters, journal articles and conference papers. She is a longstanding member of the Mathematical Association of Tasmania and served as President for 4 years. She was co-Editor of Australian primary Mathematics Classroom for several years and is currently President of the Australian Association of Mathematics Teachers.

CHEW Cheng Meng has been a mathematics educator at University Sains Malaysia in Penang since 2007. His present teaching responsibilities include mathematics education for primary and secondary teacher education students. His research interests in mathematics education include geometric thinking, the use of The Geometer's Sketchpad for teaching and learning mathematics, lesson study, and others. He has written several book chapters and papers related to geometric thinking, the use of The Geometer's Sketchpad for teaching and learning mathematics and lesson study. Chew is a member of the Association of Science and Mathematics Education Penang (ASMEP) and a member of the editorial board of SAINSAB-ONLINE: The Journal of the Association for Science and Mathematics Education.

Keiko HINO is Professor of Mathematics Education at Utsunomiya University in Japan. She received her B.Ed. and M.Ed. from Tsukuba University and Ph.D. in Education from Southern Illinois University. She began her career as a Research Assistant at Tsukuba University in 1995. After the career as Associate Professor of Mathematics Education at Nara University of Education, she is now working at Utsunomiya University. Dr. Hino has been Professor since 2010 at Utsunomiya University. Her major scholarly interests are students' development of proportional reasoning through classroom teaching, international comparative study on teaching and learning mathematics, and mathematics teachers' professional development. She has authored or co-authored 2 books, 17 book chapters, and over 40 journal articles and presented at over 40 conferences, including the International Congress on Mathematics Education, the International Conference of Psychology of Mathematics Education, the East Asia Regional Conference on Mathematics Education, and annual meetings of the Japan Society of Mathematical Education. She is also involved in activities for improving mathematics education as an editor of Japanese Primary and Lower Secondary School Mathematics Textbooks and External-expert for Lesson Study in Mathematics.

Berinderjeet KAUR is a Professor of Mathematics Education and Head of the Centre for International Comparative Studies (CICS) at the National Institute of Education in Singapore. Her primary research interests are in the area of classroom pedagogy of mathematics teachers and comparative studies in mathematics education. She has been involved in numerous international studies of Mathematics Education and is the Mathematics Consultant to TIMSS 2011. She is also a member of the MEG (Mathematics Expert Group) for PISA 2015. She is the principal investigator (Singapore) of the Learner's Perspective Study (LPS) helmed by Professor David Clarke of the University of Melbourne. As the President of the Association of Mathematics Educators (AME) from 2004-2010, she has also been actively involved in the professional development of mathematics teachers in Singapore and is the founding chairperson of the Mathematics Teachers Conferences that started in 2005. She is also the founding editor of the AME Yearbook series that started in 2009. On Singapore's 41st National Day in 2006, she was awarded the Public Administration Medal by the President of Singapore.

Marian KEMP has been a mathematics educator at Murdoch University for over 25 years, firstly in the School of Education and then the Student Learning Centre where she was the Head of the Centre from 2004. She is currently the Director of Student Life and Learning at Murdoch University. Marian has provided support for undergraduates in mathematics and statistics across the university including developing their ability to make appropriate use of scientific and graphics calculators for learning mathematics. She has published in this field and has presented papers and workshops to teachers at conferences nationally and internationally. Marian has developed programs for improving numeracy across the curriculum that have involved critical numeracy tasks, and more recently a series of online numeracy modules, for students enrolled in Murdoch's first year interdisciplinary Foundation Units. Her research has involved the development of student strategies for interpreting graphs and tables, including the use of a Five Step Framework and in this field she has published papers and presented at conferences for teachers throughout Australia and internationally. In 2007 Marian was awarded a Carrick Institute Award for University Teaching for outstanding contributions to student learning in the development of critical numeracy in tertiary curricula.

Barry KISSANE has been a mathematics educator at Murdoch University in Perth since 1985, except for a period working and studying at the University of Chicago and a recent period as the Dean of the School of Education at Murdoch University. His present teaching responsibilities include mathematics for primary teacher education students and mathematics education for secondary teacher education students. His research interests in mathematics education include numeracy, curriculum development, the use of technology for teaching and learning mathematics and statistics, popular mathematics, teacher education and others. He was written several books and many papers related to the use of graphics calculators in school mathematics, and published papers on other topics, including the use of the Internet and mathematics teacher education. Barry has served as President of the Mathematical Association of Western Australia (MAWA) and as President of the Australian Association of Mathematics Teachers (AAMT). He has been a member of editorial panels of various Australian journals for mathematics teachers for around 30 years, including several years as Editor of The Australian Mathematics Teacher. A regular contributor to conferences for mathematics teachers throughout Australasia, he is an Honorary Life member of both the AAMT and the MAWA.

Masataka KOYAMA is Professor of Mathematics Education at Hiroshima University in Japan. He received his B.Ed., M.Ed., and Ph.D. in Education from Hiroshima University. He began his career as an Assistant Professor of Mathematics Education at Hyogo University of Teacher Education in 1986 and moved back to Hiroshima University as Lecturer of Mathematics Education in 1991. Dr. Koyama has been Professor since 2007 and Vice-dean of Graduate School and Faculty of Education since 2009 at Hiroshima University. His major scholarly interests are students' mathematical understanding, international comparative study on students' mathematical attainments, mathematics teachers' professional development, and school mathematics curricula and textbooks. He has authored or co-authored 3 books, over 40 book chapters, and over 50 journal articles and presented at over 60 conferences, including the International Congress on Mathematics Education, the International Conference of Psychology of Mathematics Education, the East Asia Regional Conference on Mathematics Education, and annual meetings of the Japan Society of Mathematical Education and the Japan Academic Society of Mathematics Education.

He is also actively involved in international and national activities for improving mathematics education as Member of the Editorial Board of Educational Studies in Mathematics, Chief-editor of Japanese Primary School Mathematics Textbook, and External-expert for Lesson Study in Mathematics.

Oh Nam KWON is Professor of Mathematics Education at Seoul National University. She received her PhD in Mathematics from Indiana University in 1992. She received her MA in Mathematics from Seoul National University, her second MA in Education in Mathematics from Indiana University, and her BS from Ewha Women's University. Her earlier professional appointments include Assistant Professor and Associate Professor of Department of Mathematics Education at Ewha Women's University and Visiting Professors of Ohio State University and San Diego State University. Her research focuses on the teaching and learning of undergraduate mathematics, gender equity issues, and teacher education. She has been involved in more than 30 grants as Principal Investigator and Collaborator. She has served as committee member for numerous international (including International Programme Committee of ICME-12) and Korean organizations of mathematics education. She is serving as National Committee of Korean Institute of Curriculum and Evaluation. She received the Best Teaching Awards, Seoul National University in 2009.

JiEun LEE has earned Bachelor degree in Mathematics and a Master degree in Mathematics Education degree. Currently, she is a doctoral student of the Graduate Program in Mathematics Education of Seoul National University. She has taught middle and high school mathematics in Korea for 5 years. Her academic interests include the development of mathematical literacy for secondary mathematics students and for adults.

Chap Sam LIM gained her PhD degree from Exeter University (UK) in 1999 with a thesis on the public images of mathematics. She is currently a Professor in mathematics education at the Universiti Sains Malaysia (USM), Penang, Malaysia. She taught mathematics in secondary schools for 8 years and lectured at a teacher training college for one and half year before joining USM in 1993. She was awarded a fellowship by the Asia Scholarship Foundation (ASF) in the year 2004-5 where she spent six months in Shanghai to carry out a cross-cultural comparison study on mathematics teaching between China and Malaysia. In the year 2008-2009, she secured another award of Fulbright Scholar where she carried out another cross-cultural comparative study at the Illinois State University, USA. She has published numerous research articles, focusing on cross-cultural study, public images of mathematics, teaching mathematics in second language, and lately Lesson Study as a professional development for mathematics teachers in both international and national journals. She has initiated and promoted a number of Malaysian schools to set up Lesson Study groups since 2004. She is an active collaborator of several international research projects especially on Lesson Study and cross cultural comparative study of mathematics teaching and learning in schools. She has just co-edited and published the first book on Lesson study in Malaysia, entitled "Innovative use of GSP through lesson Study collaboration".

Auxencia Alarcon LIMJAP was a mathematics professor at the De La Salle University for 22 years. She is currently the Vice President for Academic Development at the Far Eastern University and a consultant for mathematics curriculum development of the Don Bosco System. She was vice dean of the College of Science of DLSU from 1994 until she was awarded the Fulbright-Hays Student Grant as Senior Scholar at the Ohio State University, School of Teaching and Learning in 1999. She worked with the proponent, Dr. Patricia Brosnan, for the Critical Pedagogical Rethinking project on mathematics education. As chairperson of the Science Education Department of DLSU from 2004 to 2011 she designed and offered a master degree program on teaching mathematics in basic education. She developed and conducted professional training programs for basic and tertiary mathematics teachers including a program for Nepalese master teachers. She published locally and internationally research articles on mathematics curriculum development, problem solving skills, assessment of mathematical thinking, classroom pedagogy, teacher development, educational psychology and use of technology. She has coauthored high school and college mathematics books. She was president of the Philippine Council of Mathematics Teacher Educators, MATHTED, Inc. from 2001-2004 and served it in various capacities until 2011. She contributed in national and international conferences as paper presenter and organizer.

John MASON has been teaching people mathematics since he was fifteen years old, when he started tutoring a fellow student. Over the years this developed into a lifelong interest in and concern for supporting the development of mathematical thinking in others, and especially in teachers. He developed an approach to mathematical thinking that is first and foremost experiential, having found that people learn best about others by learning about themselves through reflecting upon their own experience. Inspired by ancient psychology from the middle and far East, he formulated principles for actively researching his own practice, and offered the techniques to others as the 'discipline of noticing'. During his forty years at the Open University he was engaged in writing distance learning courses in mathematics and in mathematics education, resulting in dozens of books, pamphlets, and hundreds of professional articles and research papers. Now, as Professor Emeritus at the Open University and as Senior Research Fellow at the University of Oxford he continues to lead workshops and seminars all over the world.

TOH Tin Lam is an Associate Professor with the Mathematics and Mathematics Education Academic Group, National Institute of Education, Nanyang Technological University, Singapore. He obtained his PhD in Mathematics (Henstock-stochastic integral) from the National University of Singapore. Dr Toh continues to do research in mathematics as well as in mathematics education. He has papers published in international scientific journals in both areas. Dr Toh has taught in junior college in Singapore and was head of the mathematics department at the junior college level before he joined the National Institute of Education.

Frank Chee Tet VOON teaches medical and dental students, doctors, dentists, obstetricians, psychiatrists and surgeons at the National

University of Singapore. Following his graduation in Medicine, he obtained a PhD in Experimental Embryology. The medical and teaching background in combination with his specialist experience in Anatomy, Embryology, Information Technology, Central Administration and Psychotherapy is the basis of his interest in the physical, emotional, intellectual and spiritual growth of human beings from gestation to old age that can be enabled by the technological evolution in human culture and civilization. In recent years, with increasing knowledge being developed in the field of neurocognition, he has been delivering lectures on the mind and its cognitive processes involved in learning and memory to educators and specialists in mental health, and on the nature of mindfulness and its neuroanatomical relationship to the brain to organizations involved in spirituality.

Anne WATSON is Professor of Mathematics Education at the University of Oxford. Before becoming an academic she taught in comprehensive secondary schools in England for 13 years, pioneering activity-based learning and teaching all-attainment groups. She has a first class degree and a master degree in pure mathematics, and her doctorate was an exploration into how teachers make informal judgements about their students' mathematical learning. Her main research interests are in raising the achievement levels of previously low-attaining students, particularly those from economically-disadvantaged backgrounds; task design to scaffold higher levels of generalisation in children's thinking; qualities of teaching that bring about positive changes in children's mathematical learning. She has published numerous articles for teachers, and several books for professional audiences. Her research work is published in major academic journals in mathematics education, and she has been an invited speaker in academic contexts in all continents. Her preferred form of presentation is workshops with teachers and teacher educators because these generate two-way communication and learning. She is a Fellow of Linacre College (University of Oxford), a Fellow of the Institute of Mathematics and its Applications, and a Fellow of the International Society for Design and Development in Education. In her spare time she enjoys working on geometry problems.

WONG Khoon Yoong is an associate professor in the Mathematics and Mathematics Education Academic Group at the National Institute of Education, Nanyang Technological University, Singapore. He has worked as a mathematics educator in Australia, Brunei Darussalam, Malaysia, and Singapore. He has provided consultancy for education institutes in Chile, Hong Kong, the Philippines, and the United States in mathematics curriculum and teacher education. He has participated in the design and review of the national mathematics curriculum in Malaysia (1970s), Brunei Darussalam (1990s), and Singapore (1980s and 2000s). His research interests cover teacher education, mathematics learning strategies, use of ICT in mathematics instruction, and mathematics problem solving. His recent research projects include Teacher Education and Development Study in Mathematics (TEDS-M), Quality of Teacher Preparation in Secondary Mathematics and Science among APEC Economies, and Singapore Mathematics Assessment and Pedagogy Project (SMAPP). He is the chief editor of The Mathematics Educator (Singapore), an internationally refereed journal published by the Association of Mathematics Educators, Singapore. He holds a BSc (Hons) and Diploma in Education (both from the University of Tasmania) and PhD (University of Queensland).

WONG Oon Hua is a teaching fellow with the Mathematics and Mathematics Education Academic Group at the National Institute of Education, Nanyang Technological University. His teaching duties include pre-service modules on Teaching of Primary Mathematics I, II and III with the Degree, Diploma and Postgraduate student teachers. Prior to his attachment at NIE, he taught in the primary schools for over 20 years and had served as Head of Department for Mathematics. His interests include metacognition and the use of games and stories in learning mathematics.